Quality Risk Management in the FDA-Regulated Industry

Third Edition

José Rodríguez-Pérez

QUALITY PRESS™

Milwaukee, Wisconsin

American Society for Quality, Quality Press, Milwaukee 53203

Printed in the United States of America

28 27 26 25 24 LS 5 4 3 2 1

Publisher's Cataloging-in-Publication data

Names: Rodriguez-Perez, Jose, 1961-, author.
Title: Quality risk management in the FDA-regulated industry, third edition / by Jose Rodriguez-Perez.
Description: Includes bibliographical references. | Milwaukee, WI: Quality Press, 2024.
Identifiers: LCCN: 2024934687 | ISBN: 9781636941554 (paperback) | 9781636941561 (pdf) | 9781636941578 (epub)
Subjects: LCSH Pharmaceutical industry—Government policy—United States. | Food industry and trade—Government policy—United States. | Risk management—United States. | Total quality management—United States. | BISAC BUSINESS & ECONOMICS / Quality Control | BUSINESS & ECONOMICS / Total Quality Management | BUSINESS & ECONOMICS / Industries / Food Industry | BUSINESS & ECONOMICS / Industries / Pharmaceutical & Biotechnology
Classification: LCC HD9666.6 .R637 2024 | DDC 615.1068/4—dc23

ASQ advances individual, organizational, and community excellence worldwide through learning, quality improvement, and knowledge exchange.

Bookstores, wholesalers, schools, libraries, businesses, and organizations: Quality Press books are available at quantity discounts for bulk purchases for business, trade, or educational uses. For more information, please contact Quality Press at 800-248-1946 or books@asq.org.

To place orders or browse the selection of all Quality Press titles, visit our website at: http://www.asq.org/quality-press.

QUALITY PRESS™

Quality Press
600 N. Plankinton Ave.
Milwaukee, WI 53203-2914
Email: books@asq.org
Excellence Through Quality™

Table of Contents

APPENDICES

List of Figures and Tables

Acronyms

API—active pharmaceutical ingredient

ART—assisted reproductive technology

BLA—Biologics License Application

CAPA—corrective and preventive action

CCP—critical control point

CDC—Centers for Disease Control and Prevention

CDER—Center for Drug Evaluation and Research

CDRH—Center for Devices and Radiological Health

CFR—Code of Federal Regulations

cGMP—current good manufacturing practice

CQA—critical quality attributes

EMA—European Medicines Agency

FDA—Food and Drug Administration

FDAAA—Food and Drug Administration Amendments Act

FD&CA—Food, Drug, and Cosmetic Act

FMEA—failure mode and effects analysis

FSMA—Food Safety Modernization Act

FTA—fault tree analysis

GHTF—Global Harmonization Task Force

HACCP—hazard analysis and critical control points

HAZOP—hazard and operability analysis

ICH—International Conference on Harmonization

IEC—International Electrotechnical Commission

IMDRF—International Medical Device Regulators Forum

ISO—International Organization for Standardization

NDA—New Drug Application

OOC—out of control

OOS—out of specification

OOT—out of trend

ORA—FDA's Office of Regulatory Affairs

OTC—over the counter

PAT—process analytical technology

PDA—Parenteral Drug Association

PHA—preliminary hazard analysis

PPC—production and process control

QbD—quality by design

QMS—quality management system

QSIT—quality systems inspection technique

QSR—quality system regulation

QTPP—quality target product profile

RCA—root cause analysis

REMS—Risk Evaluation and Mitigation Strategies

RMP—risk management plan

SRP—site risk potential

USC—United States Code

Preface to the Third Edition

Since the publication of the first edition of this book in 2012, the emphasis on risk-based processes has grown exponentially across all sectors, and risk management is now considered as significant as quality management. ISO 9001, the fundamental international quality management system (QMS) standard, was recently revised under the new version, ISO 9001:2015 (International Organization for Standardization 2015), and it now requires that top management promote the use of risk-based thinking.

Another example of the greater emphasis on risk management is the international standard ISO 13485:2016, which specifies the requirements for a QMS specific to the medical device industry. The new version published in March 2016 (International Organization for Standardization 2016) shows a greater emphasis on risk management and risk-based decision-making.

A third example is the U.S. FDA Food Safety Modernization Act (FSMA), the most important reform of U.S. food safety laws in more than 70 years. It was signed into law by President Obama on January 4, 2011. It aims to ensure that the U.S. food supply is safe by shifting the focus from responding to contamination to preventing it. In other words, it is a risk-based food safety prevention system.

The purpose of this third edition of this book is to offer an updated view of the risk management field as it applies to FDA-regulated products. Three key guidelines have been recently updated [ICH Q9(R1), ISO 14971:2019, and ISO 22000:2018], and they are discussed in this new edition. This book will assist medical and food product manufacturers with the integration of a risk management system or risk management principles and activities into their existing QMS by providing practical explanations and examples. The appropriate use of quality risk management can facilitate compliance with regulatory requirements such as good manufacturing practice or good laboratory practice.

The first chapter of the book presents the historical background of the risk-based concepts and how the initial safety-only focus moved to a wider scenario with the consideration of several other elements.

The second chapter presents what is currently being done in the different areas regulated by the FDA. It describes the updated harmonization documents and international standards developed for different types of products such as medical devices, food, and drugs. This chapter now includes the risk requirement of the newest version of ISO 14971:2019 for medical devices, the updated ICH guidance for industry Q9(R1) published in May 2023, and the new version of ISO 22000:2018 with its updated requirements for food safety management systems. A new section was included to discuss the latest initiative implemented by the FDA to use risk management plans to mitigate the potential of drug shortages.

The third chapter describes the main principles of the quality risk management field, emphasizing the sound scientific basis of the quality risk management process and its close relationship with quality system elements.

The fourth chapter details the quality risk management process, defining responsibilities and then, sequentially, the whole process, starting with risk assessment, continuing with risk control and risk communication, and ending with a review and evaluation of the effectiveness of the risk process.

The fifth chapter describes the integration between risk management and quality management, including detailed examples of such integration. It also included a section covering pre- and post-market risk for medical products.

The sixth chapter presents a collection of the methodologies and tools more widely used during risk management processes. Each tool is explained in great detail, including real examples from the FDA-regulated industry. These examples make this one of the most valuable sections of this book, along with the very detailed explanation and examples of practical applications that can be found in Chapter 7, including topics such as error risk reduction, benefit–risk determination, and data integrity issues.

The eighth chapter discusses some of the most common pitfalls and misunderstandings regarding risk management, specifically those related to the use of failure mode and effects analysis (FMEA) as the only element of risk management programs.

Finally, there are five appendices, including a very detailed risk management case study (Appendix B) and an example of a quality risk management procedure (Appendix D).

Acknowledgments

Any published book is the result of many individual efforts in addition to those of the author. Many thanks to the clients of our training and consultancy business, and to many friends and colleagues for their support and for sharing their ideas and comments.

A special thank you goes to the Quality Press staff and peer reviewer Tiea Theurer for their outstanding work.

Thanks to all my associates at Business Excellence Consulting Inc. for trusting us and our idea of what a consultancy company should be.

I also need to express my deep gratitude and appreciation to Manuel Peña-Rodríguez, author of Appendices A, B, and D and coauthor of the "Fail-Safe FMEA" article included as Appendix C. Thank you for your continuous support and friendship.

1

Introduction to Quality Risk Management

WHAT IS QUALITY RISK MANAGEMENT?

Organizations of all types and sizes face internal and external factors and influences that create uncertainty about whether and when they will achieve their objectives. The effect this uncertainty has on an organization's objectives is called "risk."

All activities of an organization involve risk. Organizations manage risk by identifying it, analyzing it, and then evaluating whether the risk should be modified by risk treatment to satisfy their risk criteria. Throughout this process, they communicate and consult with stakeholders, and monitor and review the risk and the controls in place to modify the risk in order to ensure that no further risk treatment is required. All organizations manage risk to some degree, and the final goal is to integrate the process for managing risk into the organization's overall governance, strategy and planning, management, reporting processes, policies, values, and culture.

Risk management can be applied to an entire organization, throughout its many areas and levels, at any time, as well as to specific functions, projects, and activities. It's important to understand that risk management is not a one-time task. It is a continuous, never-ending process that should focus on multiple organizational aspects. All of those considerations tie into one comprehensive risk management program. Although the practice of risk management has been developed over time and within many sectors in order to meet diverse needs, the adoption of consistent processes within a comprehensive framework can help to ensure that risk is managed effectively, efficiently, and coherently across an organization.

Each specific sector or application of risk management brings with it individual needs, audiences, perceptions, and criteria. When properly implemented and maintained, the management of risk enables an organization to:

- Foster proactive management
- Improve controls and operational effectiveness and efficiency
- Be aware of the need to identify and treat risk throughout the organization
- Improve the identification of opportunities and threats
- Comply with relevant legal and regulatory requirements
- Improve mandatory and voluntary reporting
- Improve governance and stakeholder confidence and trust
- Establish a reliable basis for decision-making and planning
- Effectively allocate and use resources for risk treatment
- Improve loss prevention and minimize losses
- Improve organizational learning and flexibility

It is impossible to design and develop a medical product that is risk free. The inherently risky nature of medical products, especially those used to treat critical conditions or those that come into contact with critical systems (in the case of medical devices), means that manufacturers must thoroughly analyze their products' risks against many factors. Prior to analyzing risks, manufacturers should make a risk management plan to implement the risk management process throughout the life cycle of their medical products. This process will help manufacturers identify potential hazards and foreseeable misuse, and estimate the risks for each hazard to better control and mitigate them.

Manufacturers must use risk management to actually reduce these risks and hazards, making their products safer for users and patients.

The current management practices and processes of many organizations include components of risk management, and many organizations have already adopted a formal risk management process for particular types of risk or circumstances. The terms "risk management" and "managing risk" are commonly used. In general terms, *risk management* refers to the architecture (principles, framework, and process) for managing risks effectively, while *managing risk* refers to applying that architecture to particular risks (International Organization for Standardization 2018b).

Risk management principles are effectively utilized in many areas of business and government, including finance, insurance, occupational safety, public health, and pharmacovigilance, and by agencies regulating these industries. Although there are examples of the use of *quality risk management* in the biopharmaceutical industry today, they are limited and do not represent the full contribution that risk management has to offer. In addition, the importance of *quality systems* has been recognized in the pharmaceutical industry, and it is becoming evident that quality risk management is a valuable component of an effective quality system.

It is commonly understood that *risk* is defined as the combination of the probability of *occurrence of harm* and the *severity* of that harm. However, achieving a shared understanding of the application of risk management among diverse stakeholders is difficult because each stakeholder might perceive different potential harms, place a different probability on each harm occurring, and attribute different severities to each harm. In relation to pharmaceuticals, although there are a variety of stakeholders, including patients and medical practitioners as well as government and industry, the protection of the patient by managing the risk to quality should be considered of prime importance.

In the case of medical devices, manufacturers are generally required to have a quality management system as well as processes for addressing device-related risks. These processes for managing risk can evolve into a stand-alone management system, or manufacturers may choose to maintain these two management systems separately, but it may be advantageous to integrate them, as it could reduce costs, eliminate redundancies, and lead to a more effective management system.

ISO 14971 was developed to assist medical device manufacturers with the integration of a risk management system or risk management principles and activities into their existing quality management system by providing practical explanations and examples.

The scope of the medical device manufacturer's quality management system will define the applicability and extent of implementing risk management principles and activities.

Processes required by the quality management system and performed by suppliers to the manufacturer are the responsibility of the manufacturer. Risk management activities relating to any process within the quality management system are also ultimately the responsibility of the manufacturer.

An effective quality management system is essential for ensuring the safety and performance of medical devices. A well-defined quality management system includes safety considerations in specific areas. Given the importance of safety, it is useful to identify some key activities that specifically address safety issues and ensure appropriate input and feedback from these activities into the quality management system. The degree to which safety considerations are addressed should be commensurate with the degree of the risk and the nature of the device. Some devices present relatively low risk or have well-understood risks with established methods of risk control.

Risk management principles should be applied throughout the life cycle of medical devices and used to identify and address safety issues. In general, risk management can be characterized by phases of activities. The first phase can be the determination of levels of risk that would be acceptable in the device. Manufacturers should have a procedure or policy to determine risk acceptability criteria. These risk acceptability criteria may come from an analysis of the manufacturer's own experience with similar medical devices or research on what appear to be currently accepted risk levels by regulators, users, or patients, given the benefits derived from diagnosis or treatment with the device. Risk acceptability criteria generally should be reflective of the state of the art in controlling risks.

The second phase can be *risk analysis*. This phase starts with identifying hazards that may occur due to characteristics or properties of the device during normal use or foreseeable misuse. After hazards are identified, risks are estimated for each of the identified hazards, using available information. In the third phase, the estimated risks are compared to the risk acceptability criteria.

This comparison will determine an appropriate level of risk reduction, if necessary. This is called *risk evaluation*. The combination of risk analysis and risk evaluation is called *risk assessment*. The fourth phase can comprise risk control and monitoring activities. The manufacturer establishes actions (risk control measures) intended to eliminate or reduce each risk to meet the previously determined risk acceptability criteria. Within the limits of feasibility, one or more risk control measures may be incorporated in order to achieve this end. Risk control activities may begin as early as design input and continue through the design and development process, manufacturing, distribution, installation, and servicing, and throughout the medical device life cycle. Some regulatory schemes prescribe a fixed hierarchy of risk control measures that should be examined in the following order:

1. Inherent safety by design
2. Protective measures in the device or its manufacture
3. Information for safety, such as warnings

Throughout the life cycle of the device, the manufacturer monitors whether the risks continue to remain acceptable and whether any new hazards or risks are discovered. Information typically obtained from the quality management system—for example, production, complaints, customer feedback—should be used as part of this monitoring. If, at any time, a risk is determined to be unacceptable, the existing risk analysis should be reexamined and appropriate action *should be taken* to meet the risk acceptability criteria. If a new hazard is identified, the four phases of risk management should be performed. These activities can be performed within the framework of the quality management system.

TASTE OF RASPBERRIES, TASTE OF DEATH: THE 1937 ELIXIR SULFANILAMIDE INCIDENT

Following is a reproduction of an article by FDA's Carol Ballentine published in the June 1981 issue of the now-defunct FDA *Consumer* magazine related to the 1937 Sulfanilamide disaster.

> *By the 1930s it was widely recognized that the Food and Drugs Act of 1906 was obsolete, but bitter disagreement arose as to what should replace it. By 1937 most of the arguments had been resolved but Congressional action was stalled. Then came a shocking development—the deaths of more than 100 people after using a drug that was clearly unsafe. The incident hastened final enactment in 1938 of the Federal Food, Drug, and Cosmetic Act, the statute that today remains the basis for FDA regulation of these products.*
>
> "Nobody but Almighty God and I can know what I have been through these past few days. I have been familiar with death in the years since I received my M.D. from Tulane University School of Medicine with the rest of my class of 1911. Covington County has been my home. I have practiced here for years. Any doctor who has practiced more than a quarter of a century has seen his share of death.
>
> But to realize that six human beings, all of them my patients, one of them my best friend, are dead because they took medicine that I prescribed for them innocently, and to realize that that medicine which I had used for years in such cases suddenly had become a deadly poison in its newest and most modern form, as recommended by a great and reputable pharmaceutical firm

in Tennessee: well, that realization has given me such days and nights of mental and spiritual agony as I did not believe a human being could undergo and survive. I have known hours when death for me would be a welcome relief from this agony." (Letter by Dr. A.S. Calhoun, October 22, 1937)

The medicine that killed Dr. Calhoun's patients was Elixir Sulfanilamide. During September and October 1937, this drug was responsible for the deaths of more than 100 people in 15 states, as far east as Virginia and as far west as California. The drug and the deaths led to the passage of the 1938 Food, Drug, and Cosmetic Act, which increased FDA's authority to regulate drugs.

Sulfanilamide, a drug used to treat streptococcal infections, had been shown to have dramatic curative effects and had been used safely for some time in tablet and powder form. In June 1937, however, a salesman for the S. E. Massengill Co., in Bristol, Tenn., reported a demand in the southern states for the drug in liquid form. The company's chief chemist and pharmacist, Harold Cole Watkins, experimented and found that sulfanilamide would dissolve in diethylene glycol. The company control lab tested the mixture for flavor, appearance, and fragrance and found it satisfactory. Immediately, the company compounded a quantity of the elixir and sent shipments—633 of them—all over the country.

The new formulation had not been tested for toxicity. At the time the food and drugs law did not require that safety studies be done on new drugs. Selling toxic drugs was, undoubtedly, bad for business and could damage a firm's reputation, but it was not illegal.

Because no pharmacological studies had been done on the new sulfanilamide preparation, Watkins failed to note one characteristic of the solution. Diethylene glycol, a chemical normally used as antifreeze, is a deadly poison. The first shipments were sent out in early September. On October 11, the American Medical Association (AMA) received reports from physicians in Tulsa, Okla., that an unfamiliar sulfanilamide compound was responsible for a number of deaths. The AMA asked for samples of the drug and then wired the Massengill Co., requesting the composition of the compound. The AMA laboratory isolated diethylene glycol as the toxic ingredient and immediately issued a warning, through newspapers and radio, that Elixir Sulfanilamide was toxic and deadly.

A New York physician learned of the deaths on October 14 and promptly notified Food and Drug Administration headquarters. An inspector from the agency's Kansas City station confirmed that eight children and one adult had died and that all had taken a product labeled "Elixir Sulfanilamide, the S. E. Massengill Co., Manufacturing Pharmacists, Bristol, Tenn.–Va."

Inspectors were immediately dispatched to the firm's headquarters in Bristol and to branch offices in Kansas City, New York, and San Francisco. They found that the firm had already learned of the poisonous effects of the liquid sulfanilamide and had sent telegrams to more than 1,000 salesmen, druggists, and doctors. However, the telegrams merely requested the return of the product and failed to indicate the urgency of the situation or say the drug was lethal. At FDA's insistence, the firm sent out a second wave of messages, worded more strongly: "Imperative you take up immediately all elixir sulfanilamide dispensed. Product may be dangerous to life. Return all stocks, our expense."

FDA then set out to make sure all the drug was retrieved. Practically the entire field force of 239 FDA inspectors and chemists was assigned to the task. State and local health officials joined the search. Newspapers and radio stations continued to issue warnings. The staff began by checking the company's shipping records and the distribution lists in the four distributing houses and in a number of wholesale and retail drugstores. Thousands of order slips were examined one by one. In one establishment alone, 20,000 sales slips were checked.

FDA employees tracked down the firm's 200 salesmen and questioned them about the dispersion of shipments and physician samples. Finding the salesmen was the first problem. In one typical case, a salesman was reported to be in a hotel in Washington, D.C. He was not there, but forwarding addresses had been left for him in Jackson, Mich., and in Baltimore. These turned out to be for another man with the same name. Four days of searching finally found the man in University Park, Md. Once the salesmen were found, there was still the problem of getting the distribution information. One man in Texas, for instance, revealed the necessary information only after being jailed by state authorities.

In many cases, locating the purchasers of the elixir required some real detective work. In some drugstores, the elixir had been sold without prescriptions to purchasers whose names the druggist didn't know. In other cases, doctors had incomplete records—or none at all—of the names and addresses of patients for whom they had prescribed. In East St. Louis, Ill., for instance, 49 prescriptions were filled and the only identification on some were such notations as "Betty Jane, 9 months old," or "Mrs. Jackson (no address)."

Even when the purchaser was finally located, the inspectors frequently needed to do some ingenious questioning to determine what happened to the purchased elixir. One East St. Louis woman told an inspector she had destroyed the drug. The inspector persisted in his questions, however. What did she mean "destroy"? Had she poured it down the sink? Had she buried it? No, the woman said, she had thrown the bottle out the window into an alley. The inspector found the bottle still unbroken, still containing enough elixir to kill any child intrigued enough to swallow its sweet, raspberry-flavored contents.

Many doctors and pharmacists did everything in their power to recover the elixir. One physician postponed his wedding to help an FDA chemist search for a three-year-old boy whose family had moved into mountain country after obtaining a prescription.

Other physicians apparently were reluctant to admit that they had prescribed the drug, perhaps fearing they would be held accountable for its consequences. One inspector, checking out a Georgia drugstore, was told that a shipment of one gallon of elixir had been returned to the manufacturer after only 6 ounces had been dispensed. The patient who had taken the 6 ounces had suffered no ill effects, the druggist reported, and the inspector confirmed that this was true. But the inspector assigned to Bristol reported that 12 ounces was actually missing from the returned gallon, so the inspector in Georgia did some more questioning around town and tuned his ears to the local gossip. He learned that two other people had also bought the elixir. Both had died.

Similarly, a South Carolina doctor told an inspector that he had dispensed the medicine to only five people and none had died. But when the inspector began asking questions around town, someone told him of the death of a lumber mill employee. The inspector recognized the symptoms as those characteristic of Elixir Sulfanilamide poisoning. Through the mill superintendent, he located the employee's sister. Yes, she said, her brother had gone to the doctor and had been given some red medicine before he died. She told the inspector that, in accordance with custom, all medicines and sickroom utensils had been placed on the grave, about one and a half miles back in the fields. Accompanied by family and friends, the dead man's sister and the inspector walked to the wooded knoll. On the single mound of fresh earth were several bottles, dishes, spoons, and a 4-ounce bottle containing about 1 ounce of Elixir of Sulfanilamide. It bore the weathered but legible prescription label of the doctor. In fact, the inspector learned, four of the doctor's patients had died after taking the elixir.

Victims of Elixir Sulfanilamide poisoning—many of them children being treated for sore throats—were ill about 7 to 21 days. All exhibited similar symptoms, characteristic of kidney failure: stoppage of urine, severe abdominal pain, nausea, vomiting, stupor, and convulsions. They suffered intense and unrelenting pain. At the time there was no known antidote or treatment for diethylene glycol poisoning.

In a letter to President Franklin D. Roosevelt, a woman described the death of her child:

"The first time I ever had occasion to call in a doctor for [Joan] and she was given Elixir of Sulfanilamide. All that is left to us is the caring for her little grave. Even the memory of her is mixed with sorrow for we can see her little body tossing to and fro and hear that little voice screaming with pain and it seems as though it would drive me insane.... It is my plea that you will take steps to prevent such sales of drugs that will take little lives and leave such suffering behind and such a bleak outlook on the future as I have tonight."

A few simple tests on experimental animals would have demonstrated the lethal properties of the elixir. Even a review of the current existing scientific literature would have shown that other studies—such as those reported in several medical journals—had indicated that diethylene glycol was toxic and could cause kidney damage or failure. But in 1937 the law did not prohibit the sale of dangerous, untested, or poisonous drugs. Dr. Samuel Evans Massengill, the firm's owner, said: "My chemists and I deeply regret the fatal results, but there was no error in the manufacture of the product. We have been supplying a legitimate professional demand and not once could have foreseen the unlooked-for results. I do not feel that there was any responsibility on our part." The firm's chemist apparently did not share this feeling; Harold Watkins committed suicide after learning of the effects of his latest concoction.

Through the dogged persistence of federal, state, and local health agencies and the effects of the AMA and the news media, most of the elixir was recovered. Of 240 gallons manufactured and distributed, 234 gallons and 1 pint were retrieved; the remainder was consumed and caused the deaths of the victims.

Twenty-five seizures were made under federal law. The charge was misbranding. "Elixir," FDA said, implied the product was an alcoholic solution, whereas it was, in fact, a diethylene glycol solution and contained no alcohol. If the product had been called a "solution" instead of an "elixir," no charge of violating the law could have been made. FDA would have had no legal authority to ensure the recovery of the drug, and many more people probably would have died.

FDA Commissioner Walter Campbell, who was then pressing for better federal regulation of drugs, pointed out how the inadequacy of the law had contributed to the disaster. "It is unfortunate that under the terms of our present inadequate Federal law, the Food and Drug Administration is obliged to proceed against this product on a technical and trivial charge of misbranding.... [The Elixir Sulfanilamide incident] emphasizes how essential it is to public welfare that the distribution of highly potent drugs should be controlled by an adequate Federal Food and Drug law. ... We should not lose sight of the fact that we had many deaths and cases of blindness resulting from the use of another new drug, dinitrophenol, which was recklessly placed upon the market some years ago. Deaths and blindness from this [drug] are continuing today. We also should remember the deaths resulting from damage to the liver that have occurred from cinchophen poisoning, a drug often recommended in such painful conditions as rheumatism. We also have unfortunate poisoning, acute and chronic, resulting from thyroid and radium preparations improperly administered to the public.

"These unfortunate occurrences may be expected to continue because new and relatively untried drug preparations are being manufactured almost daily at the whim of the individual manufacturer, and the damage to public health cannot accurately be estimated. The only remedy for such a situation is the enactment by Congress of an adequate and comprehensive national Food and Drugs Act which will require that all medicines placed upon the market shall be safe to use under the directions for use...."

As it turned out, the elixir experience did more than hasten enactment of the 1938 Federal Food, Drug, and Cosmetic Act.

The New Drug section, added to prevent such tragedies, gave the United States a new system of drug control, which provided superior protection while stimulating medical research and progress. And 25 years later, it saved the nation from an even greater drug tragedy—a thalidomide disaster—like that in Germany and England. Here again, history repeated itself. A pending bill, the Drug Amendments of 1962, was finally enacted.

SEVENTY YEARS LATER

At the end of September 2006, several hospitals in Panama City (Panama) were filling up with patients with contradictory symptoms. The symptoms seemed to match with Guillain–Barré syndrome, but these patients were also losing their ability to urinate, a symptom not related to Guillain-Barré. The death rate of this mysterious illness was nearly 50 percent when hospital management decided to isolate all the patients with the illness in a large room so doctors could compare notes and theories. Soon, patients from other parts of the country started to arrive at hospitals. Doctors had no idea what was happening: the mysterious illness was attacking elderly citizens with hypertension and diabetes history. About half were receiving a blood pressure medicine, and many did not remember clearly if they had been taking other drugs. Suspecting something wrong with the hypertension drug, the medicine was removed from the pharmacies while the U.S. FDA conducted laboratory analyses, revealing that the blood pressure drug was safe.

Most of the patients taking the blood pressure medicine developed a cough (a common side effect), and they were prescribed an expectorant. Immediately, biological samples and the syrup were sent to the United States for analysis. When urine analyses for a series of metals, pesticides, or their metabolites resulted negative, Centers for Disease Control (CDC) scientists recalled previous incidents in Nigeria and Haiti. The CDC analysis revealed that the samples contained approximately 8 percent v/v diethylene glycol (DEG). Later, raw glycerin was analyzed and results revealed 22.2 percent v/v DEG.

The Panamanian government implemented a nationwide campaign, collecting around 6,000 bottles of cough syrup and three other products with the tainted glycerin manufactured by Social Security Laboratories. The 46 barrels of syrup were bought by Social Security Laboratories through a Panamanian middleman, Grupo Comercial Medicom, who bought the product from Rasfer Internacional, a Spanish company. In fact, Rasfer received the product from CNSC Fortune Way, which in turn bought it from the Taixing Glycerine Factory. At the request of the

United States, the Chinese FDA investigated the Taixing Glycerine Factory and CNSC Fortune Way, but the agency concluded that it was not under their jurisdiction because the factory was not certified to make medicine. Taixing sold the syrup as "TD glycerin," wording that neither Chinese inspectors nor Spanish medical authorities comprehended. Unfortunately, Taixing used "TD" for the Chinese word *tidai*, meaning "substitute."

The New York Times reported that Taixing was closed by the Chinese government and CNSC Fortune Way was never sanctioned (Bogdanich and Hooker 2007). In Spain, Rasfer Internacional declared bankruptcy after the lawyer of Medicom filed a lawsuit for $400 million in July 2008. Spanish authorities prosecuted the general manager of Rasfer Internacional. Meanwhile, in Panama, the legal representative of Medicom was jailed pending trial. Seventeen other persons were also charged in relation to the incident.

The Panama case caused CDC to set standardized methodology for DEG identification, hoping to have a more timely response in future events. After that incident, FDA also issued an Industry Guidance document "intended to alert pharmaceutical manufacturers, pharmacy compounders, repackers, and suppliers to the potential public health hazard of glycerin contaminated with DEG and recommended appropriate testing procedures for the use of glycerin." As of June of 2011, the number of confirmed deaths, according to the official Panama government list, had risen to 219 victims.

In 2022, more than 300 children (mainly younger than 5 years old) in Gambia, Indonesia, and Uzbekistan died of acute kidney injury, in deaths that were associated with contaminated medicines (over-the-counter cough syrups) having high levels of diethylene glycol and ethylene glycol.

RISK MANAGEMENT AND RISK ASSESSMENT

Risk management is defined in ISO 14971:2019 as "the systematic application of management policies, procedures, and practices to the task of analyzing, evaluating, controlling, and monitoring risk." As such, risk management is an ongoing, continuous, and iterative process.

Risk assessment can be considered as a subset of the risk management process; it is the first step in the risk management process, and it involves both the analysis and evaluation of risks. The complete risk management process, in addition to risk assessment, includes risk control and risk monitoring. Chapter 4 details the different elements of the risk management process.

RISK-BASED APPROACH

Developing a risk culture is one of the most important tasks a regulated firm can do to manage risk. All medical products carry some degree of risk (from design and manufacturing errors to use errors), and even the safer, over-the-counter medicines and devices are highly regulated. Regulators, by definition, are inherently risk averse, and FDA is no exception; however, in 2002, it announced a significant new initiative, "Pharmaceutical cGMPs for the 21st Century: A Risk-Based Approach," to enhance and modernize the regulation of pharmaceutical manufacturing and product quality—to bring a twenty-first-century focus to this critical FDA responsibility. The initiative was intended to modernize the FDA's regulation of pharmaceutical quality for veterinary and human drugs and select human biological products such as vaccines.

As part of this initiative, both the pharmaceutical and the chemistry, manufacturing, and controls (CMC) regulatory programs were evaluated with the following objectives in mind:

- Encourage the early adoption of new technological advances by the pharmaceutical industry
- Facilitate industry application of modern quality management techniques, including implementation of quality systems approaches, to all aspects of pharmaceutical production and quality assurance
- Encourage implementation of risk-based approaches that focus both industry and agency attention on critical areas
- Ensure regulatory review, compliance, and inspection policies are based on state-of-the-art pharmaceutical science
- Enhance the consistency and coordination of FDA's drug quality regulatory programs, in part by further integrating enhanced quality systems approaches into the Agency's business processes and regulatory policies concerning review and inspection activities

Early in the initiative, a number of multidisciplinary working groups were formed, comprising FDA experts from various areas of scientific and regulatory practice within FDA. As a result of this work, FDA completed its assessment of the existing cGMP programs and began to develop the product quality regulatory system of the future, following these guiding principles, in order to minimize the risks to the public health associated with pharmaceutical product manufacturing:

- Risk-based orientation
- Science-based policies and standards
- Integrated quality systems orientation
- International cooperation
- Strong public health protection

FDA identified a risk-based orientation as one of the driving principles of the cGMP initiative. The progress outlined below reflects FDA's commitment to the adoption of risk management principles that will enhance the Agency's inspection and enforcement program, which is focused on protecting the public health.

FDA's Strategic Action Plan

The Agency's strategic plan identified efficient risk management as a key element. Efficient risk management requires using the best scientific data, developing quality standards, and using efficient systems and practices that provide clear and consistent decisions and communications for the American public and regulated industry. FDA identified efficient risk management as the primary way to make the most effective use of Agency resources and address these challenges. This approach incorporates rigorous analysis to consistently identify the most important risks, and the use of a quality systems approach to designing, conducting, and evaluating FDA core business processes.

Risk-Based Model for Inspectional Oversight

The Agency also developed a risk-based model for prioritizing sites for manufacturing inspections, which is discussed in Chapter 5.

Part 11 Guidance

With the issuance in 2003 of the guidance for industry Part 11, *Electronic Records, Electronic Signatures—Scope and Application,* many barriers to scientific and technological advances were removed, and the use of risk-based approaches to managing computer systems is encouraged.

Aseptic Processing Guidance

The final guidance for industry published in 2004, *Sterile Drug Products Produced by Aseptic Processing—Current Good Manufacturing Process,* advocates a risk-based framework, underscoring the value of proactive approaches to ensuring sterility. Among the sections of the guidance that address risk-based approaches are those relating to key roles played by personnel, design, environmental control, and media fills in an aseptic processing operation. This guidance states that "nearly all drugs recalled due to non-sterility or lack of sterility assurance in the period spanning 1980–2000 were produced via aseptic processing." (Food and Drug Administration 2004a)

In February 2023, FDA warned consumers and healthcare practitioners not to purchase and also to immediately stop using two brands of artificial tears due to potential bacterial contamination. These were over-the-counter products, manufactured by Global Pharma Healthcare Private Limited, intended to be sterile. Global Pharma initiated a voluntary recall at the consumer level of all unexpired lots of EzriCare Artificial Tears and Delsam Pharma's Artificial Tears. FDA recommended this recall due to the company's current good manufacturing practice (cGMP) violations, including lack of appropriate microbial testing, formulation issues (the company manufactures and distributes ophthalmic drugs in multi-use bottles, without an adequate preservative), and lack of proper controls concerning tamper-evident packaging.

As of January 31, 2023, the CDC identified 55 patients in 12 states with infections that have been linked by epidemiologic and laboratory evidence to the use of EzriCare Artificial Tears. Associated adverse events include hospitalization, one death with bloodstream infection, and permanent vision loss from eye infections. The CDC issued an alert recommending consumers stop using EzriCare Artificial Tears pending additional guidance from the CDC and FDA.

FDA also placed Global Pharma Healthcare Private Limited on import alert for providing an inadequate response to a records request and for not complying with cGMP requirements. The import alert prevents these products from entering the United States. As of May 15, 2023, CDC identified 81 patients in 18 states with a rare strain of extensively drug-resistant *Pseudomonas aeruginosa*. Fourteen patients had vision loss, an additional four patients required enucleation (surgical removal of eyeball), and four died.

MORE THAN SAFETY RISKS: FROM SAFETY RISKS TO QUALITY BY DESIGN

An effective risk management process involves many complex and interrelated activities whose final goal is to *protect the patient* and users of medical products, which is exactly the mission of FDA.

When properly managed, the risk management process will help regulated companies ensure customers' and users' safety, and protect company value and reputation. On the other hand, when ineffective risk management processes are in place, some consequences can be:

- Injuries and even deaths related to unsafe medical products
- Regulatory actions, including costly product recalls, warning letters, and even injunctions
- Tarnished corporate image
- Decreased value for stakeholders and stockholders

Better design of medical products, including the design of their manufacturing and quality control processes, is a critical element of the risk management process. However, there is another crucial phase, which is during the use of the product. To minimize this type of risk (for example, an error when using a medical device or taking a drug), FDA developed specific guidances, among them:

- *Risk Evaluation and Mitigation Strategies* (REMS), which require medication guides for some drug and biological products
- *Quality by Design,* which aims to design safer and better drugs and biotechnology products
- *Design Control Guidance for Medical Device Manufacturers,* which is focused on meeting user needs
- *Do It by Design: An Introduction to Human Factors in Medical Devices*

Medication Guides

Medication guides were introduced in the late 1990s for products perceived to pose "serious and significant public health concerns." It was mandatory for pharmacists to distribute them at the time of drug dispensing. For some medical products (mainly drug and biological products) that

FDA determines pose a serious and significant public health concern, distribution of FDA-approved patient medication information that is necessary to patients' safe and effective use of the drug products is required. It is called a medication guide, and it applies to products used primarily on an outpatient basis without direct supervision by a healthcare professional. These regulations are codified in 21 CFR part 208.

Quality by Design

Quality by design (QbD) is a concept first outlined by well-known quality expert Joseph M. Juran in various publications, most notably *Juran on Quality by Design*. Juran believed that quality could be planned, and that most quality crises and problems relate to the way in which quality was planned in the first place. QbD is a systematic approach to the development of new products and processes, and it begins with predefined objectives, emphasizes product and process understanding and process control, and is based on sound science and quality risk management.

The concept promotes industry's understanding of the product and manufacturing process, starting with product development, basically building quality in, not testing it in. Under this concept, when designing and developing a product, a company needs to define desired product performance and identify its *critical quality attributes* (CQAs). Based on this information, the company then designs the product formulation and process to meet these product attributes.

This leads to an understanding of the impact of raw material attributes and process parameters on the CQAs, and identification and control of sources of variability. As a result of all this knowledge, a company can continually monitor and update its manufacturing process to assure consistent product quality. This systematic approach to product development and manufacturing varies a great deal from the traditional approach, which was extremely empirical.

While QbD principles have been used to advance product and process quality in every industry, particularly the automotive industry, they have most recently been adopted by FDA as a vehicle for the transformation of how drugs are discovered, developed, and commercially manufactured. During an FDA presentation at the 2007 PDA/FDA Joint Conference 2007, the Agency presented the following benefits from the adequate implementation of QbD:

- Enhancing the scientific foundation for review
- Providing for better coordination across review, compliance, and inspection

- Improving information in regulatory submissions
- Providing for better consistency
- Improving the quality of preapproval review
- Providing for more flexibility in decision-making
- Ensuring decisions are made based on science and not on empirical information
- Involving various disciplines in decision-making
- Using resources to address higher risks

Benefits for the industry include:

- Ensuring better design of products with fewer problems in manufacturing
- Reducing the number of manufacturing supplements required for post-market changes, and fostering reliance on process and risk understanding and risk mitigation
- Allowing for implementation of new technology to improve manufacturing without regulatory scrutiny
- Allowing for possible reduction in overall costs of manufacturing (less waste)
- Ensuring less hassle during review (fewer deficiencies, quicker approvals)
- Improving interaction with FDA—dealing on a science level instead of on a process level
- Allowing for continuous improvements in products and manufacturing processes
- Allowing for better understanding of how active pharmaceutical ingredients (APIs) and excipients affect manufacturing
- Relating manufacturing to clinical considerations during design
- Providing a better overall business model

Design Controls for Medical Devices

Risk management is the systematic application of management policies, procedures, and practices to the tasks of identifying, analyzing, controlling, and monitoring risk. It is intended to be a framework within which experience, insight, and judgment are applied to successfully manage risk. It is included in this guidance because of its effect on the design process.

Risk management begins with the development of the design input requirements. As the design evolves, new risks may become evident. To systematically identify and, when necessary, reduce these risks, the risk management process is integrated into the design process. In this way, unacceptable risks can be identified and managed earlier in the design process, when changes are easier to make and less costly.

An example of this is an exposure control system for a general-purpose X-ray system. The control function was allocated to software. Late in the development process, risk analysis of the system uncovered several failure modes that could result in overexposure to the patient. Because the problem was not identified until the design was near completion, an expensive, independent backup timer had to be added to monitor exposure times.

The Safe Medical Devices Act of 1990 added design validation requirements to the cGMP requirements in section 520(f) of the Act. Section 820.30 of the Quality System Regulation (QSR) lists the design control requirements that manufacturers should satisfy to be in compliance.

Design Control Guidance for Medical Device Manufacturers is another document that may assist manufacturers in understanding the intent of the design control requirements. This manual interprets the language of the QSR and explains the underlying concepts in practical terms.

Do It by Design: An Introduction to Human Factors in Medical Devices is a document that contains background information about human factors as a discipline, describes and illustrates device problems, and discusses human factors principles and methods as a part of the design control system. Both of these manuals are possible resources for manufacturers that are either developing or improving their design control system.

The design controls section 820.30 of the QSR applies to the design of products and processes, and changes to existing designs and processes. Changes to existing designs should be made in accordance with design control requirements even if the original design was not subject to these requirements. Each manufacturer of any class III or class II device, and a few categories of class I devices, including automated devices with computer software, shall establish and maintain procedures to control the design of the device to ensure certain specified design requirements are met.

ISO 31000

The international standard ISO 31000 is intended to be a family of standards relating to risk management codified by the International Organization for Standardization (ISO). The purpose of ISO 31000:2018 is to provide principles and generic guidelines on risk management. ISO 31000 seeks to provide a universally recognized paradigm for practitioners and companies employing risk management processes to replace the myriad of existing standards, methodologies, and paradigms that differ between industries and regions, and by subject matter.

The purpose of ISO 31000:2018 is to be applicable and adaptable for "any public, private or community enterprise, association, group or individual." Accordingly, the general scope of ISO 31000—as a family of risk management standards—was not developed with a particular industry group, management system, or subject matter field in mind, but rather to provide best practice structure and guidance to all operations concerned with risk management.

ISO 31000:2018 provides principles and generic guidelines on risk management, and it can be applied throughout the life of an organization and to a wide range of activities, including strategies and decisions, operations, processes, functions, projects, products, services, and assets. ISO 31000:2018 can be applied to any type of risk, whatever its nature, whether having positive or negative consequences.

Although ISO 31000:2018 provides generic guidelines, it is not intended to promote uniformity of risk management across organizations. The design and implementation of risk management plans and frameworks should take into account the varying needs of a specific organization, its particular objectives, context, structure, operations, processes, functions, projects, products, services, or assets, and specific practices employed. It is intended that ISO 31000:2018 be utilized to harmonize risk management processes in existing and future standards. It provides a common approach in support of standards dealing with specific risks and/or sectors, but does not replace those standards. ISO 31000:2018 is not intended for the purpose of certification.

Currently, the ISO 31000 family includes:

- ISO 31000:2018 *Risk management—Guidelines*
- IEC 31010:2019 *Risk management—Risk assessment techniques*
- ISO 31073:2022 *Risk management—Vocabulary*

HISTORICAL PERSPECTIVE AND CURRENT SITUATION

The roots of modern risk management can be tied to a cohort of regulatory initiatives in the latter part of the past century. The decision to approve a drug or any medical product (and its classification as prescription or nonprescription) is highly dependent on the answer to the following question: "Do the benefits of the medical product outweigh the risks?" Both manufacturers and FDA continually evaluate the potential benefits of a medical product versus its potential risks. Many factors are considered, some of which include:

- Seriousness of the disease or condition to be treated
- Size of the patient population
- Expected benefit of the medical product
- Expected duration of treatment
- Seriousness of the known or potential adverse events

These evaluations are performed not only prior to the approval of a new medical product but also throughout the entire life cycle of the product. This serves as a means to continuously assess the safety and efficacy of existing product based on adverse event reports and results from post-marketing clinical studies.

In 1960, FDA instituted a policy of full disclosure. Manufacturers were required (and still are) to provide complete information about the product's indication, dosing, side effects, and so on, to healthcare professionals. This information is included in the product labeling.

The Controlled Substances Act (CSA) of 1970 is another early example of risk management. The CSA regulates manufacturers, prescribers, and dispensers as well as the label and warning text, providing additional indicators of risk and tools to manage those risks beyond those established under the Federal Food, Drug, and Cosmetic Act (FD&C Act). Common tools included boxed warnings and "Dear Healthcare Provider" letters.

The safe and effective use of drugs was the reason FDA endorsed the requirement in the late 1970s to distribute patient package inserts (PPIs) with oral contraceptives and estrogen replacements to patients. Prior safety communications were geared toward healthcare professionals. Since oral contraceptives and estrogens were considered elective medications, FDA insisted that patients (consumers) be provided with written information about the benefits and risks of taking these products to assist in their decision about birth control methods.

Special programs that extended beyond labeling and PPIs to restrict access to certain risky products were instituted between 1988 and 1998. Some products (Accutane, Clozaril, and thalidomide, for example) had specifically designed programs to detect or prevent serious adverse events. In the case of thalidomide, used to treat complications of leprosy, it was approved in the United States (1998) with The System for Thalidomide Education and Prescribing Safety (S.T.E.P.S.). The program required that all patients, pharmacists, and physicians be registered, and that patients be tested for pregnancy to avoid the potential for birth defects.

In the late 1990s, after the removal of a number of high-profile drugs from the market due to emergent safety concerns (for example, fenfluramine), FDA's efforts with respect to drug safety, both pre- and post-marketing, came under scrutiny. Questions arose about the pre-market and post-market assessments of safety and drug risks and whether the shorter review time frames specified by the Prescription Drug User Fee Act of 1992 (PDUFA) were resulting in the approval of more-dangerous drugs. As one of her first initiatives after being sworn in as FDA Commissioner in 1998, Dr. Jane Henney established a task force to evaluate the system for managing the risks of FDA-approved medical products, focusing particularly on FDA's part in the system. The result of that review was published in May 1999 (*Managing the Risks from Medical Product Use: Creating a Risk Management Framework*, report to the FDA commissioner from the task force on risk management). This report heralded the modern risk management era (Food and Drug Administration 1999b).

The task force assessed risk management practices within the overall healthcare delivery system, focusing on the roles and responsibilities of each participant. The task force applied a risk management model used in other federal sectors. They also examined the various risks from medical products and their sources. The task force then evaluated FDA's role in the current system. First, they reviewed the Agency's *pre-marketing* risk assessment and approval processes to determine if serious adverse events were occurring at a higher rate than they had in the past. Next, the task force evaluated FDA's *post-marketing* surveillance and risk assessment programs to see if they were doing the job they were intended to do. Finally, the task force analyzed all of FDA's risk management activities to evaluate the Agency's role in the overall system for managing medical product risks. Medical products provide great benefit to the public, but they can also cause injury.

FDA and the many other participants in healthcare delivery act to maximize the benefits and minimize the risks associated with using medical products, but often the actions of the participants are insufficiently integrated. The task force believed that the common goal of maximizing benefits and minimizing risks could be greatly advanced if the participants in the system worked together to gain an understanding of these activities within a systems framework. To achieve such a framework, it is necessary to gain a better understanding of the risks involved and their sources, and it is also necessary to clarify the role of each stakeholder and ensure that these individual roles are well integrated. Only then can effective risk management strategies be planned.

In July 2006 the Parenteral Drug Association's (PDA) Risk Management Task Force for Aseptic Processes conducted an electronic survey among its members to determine current industry practices regarding implementation of quality risk management in their organizations. This electronic survey was open and publicly available via the PDA website and targeted professionals in the regulated industry (pharmaceutical, medical devices, and biotechnology manufacturers) who are involved in initiating, implementing, or reviewing risk management programs or decisions in their organizations. One hundred twenty-nine members participated (Ahmed et al. 2008).

Among the major findings were:

- The "aseptic processing/filling" operation is the functional area identified as having the greatest need for risk assessment and quality risk management.
- The most widely used methodology in industry to identify risk is failure mode and effects analysis (FMEA). This tool is most widely applied in assessing change control and for adverse event, complaint, or failure investigations.
- Despite the fact that personnel training was identified as the strategy most used for controlling/minimizing risk, the largest contributors to sterility failure in operations are still "personnel."
- Most companies still rely on "manufacturing controls" to mitigate risk and deem the utilization of process analytical technology (PAT) least important in this aspect.
- A majority of correspondents verified that they did not periodically assess their risk management programs.
- A majority of the correspondents desired to see case studies or examples of risk analysis implementation (as applicable to aseptic processing) in future PDA technical reports on risk management.

ISO 9001:2015 RISK REQUIREMENTS

ISO 9001 is the international standard that specifies requirements for a quality management system (QMS). Organizations use the standard to demonstrate their ability to consistently provide products and services that meet customer and regulatory requirements. It is the most popular standard in the ISO 9000 series and the only standard in the series to which organizations can certify.

ISO 9001 was first published in 1987 by the International Organization for Standardization (ISO), an international agency composed of the national standards bodies of more than 160 countries. The current version (fifth edition) of ISO 9001 was released in September 2015.

There are many elements of risk-based thinking in the ISO 9001:2015 international standard that may affect organizations as they work toward compliance to the standard. The following excerpts and summaries describe references to risk in ISO 9001:2015 (International Organization for Standardization 2015):

Under Introduction, Section 0.1 General, the standard establishes that "the adoption of a quality management system is a strategic decision for an organization that can help to improve its overall performance and provide a sound basis for sustainable development initiatives." And it cites as one of the main potential benefits of implementing a QMS based on this international standard the addressing of risks and opportunities associated with its context and objectives.

This international standard employs the process approach, which incorporates the plan–do–check–act (PDCA) cycle and risk-based thinking. Risk-based thinking enables an organization to determine the factors that could cause its processes and its QMS to deviate from the planned results, to put in place preventive controls to minimize negative effects, and to make maximum use of opportunities as they arise.

Under Section 0.3 Process approach, 0.3.1 General, this international standard establishes that "the process approach involves the systematic definition and management of processes, and their interactions, so as to achieve the intended results in accordance with the quality policy and strategic direction of the organization. Management of the processes and the system as a whole can be achieved using the PDCA cycle with an overall focus on risk-based thinking aimed at taking advantage of opportunities and preventing undesirable results."

The *plan* element of the PDCA cycle establishes the objectives of the system and its processes, and the resources needed to deliver results in accordance with customers' requirements and the organization's policies and identify and address risks and opportunities.

In Section 0.3.3 Risk-based thinking, this international standard establishes that risk-based thinking:

> is essential for achieving an effective quality management system. The concept of risk-based thinking has been implicit in previous editions of this International Standard including, for example, carrying out preventive action to eliminate potential nonconformities, analyzing any nonconformities that do occur, and taking action to prevent recurrence that is appropriate for the effects of the nonconformity.
>
> To conform to the requirements of this International Standard, an organization needs to plan and implement actions to address risks and opportunities. Addressing both risks and opportunities establishes a basis for increasing the effectiveness of the quality management system, achieving improved results and preventing negative effects.
>
> Opportunities can arise as a result of a situation favorable to achieving an intended result, for example, a set of circumstances that allow the organization to attract customers, develop new products and services, reduce waste or improve productivity. Actions to address opportunities can also include consideration of associated risks. Risk is the effect of uncertainty and any such uncertainty can have positive or negative effects. A positive deviation arising from a risk can provide an opportunity, but not all positive effects of risk result in opportunities.

It declares that this international standard enables an organization to use the process approach, coupled with the PDCA cycle and risk-based thinking, to align or integrate its QMS with the requirements of other management system standards.

Under Section 4.4 Quality management system and its processes, Subsection 4.4.1, it establishes that "The organization shall address the risks and opportunities as determined in accordance with the requirements of section 6.1."

Under Section 5.1 Leadership and commitment, Subsection 5.1.1 General, it establishes that "Top management shall demonstrate leadership and commitment with respect to the quality management system by promoting the use of the process approach and risk-based thinking."

Under Section 5.1.2. Customer focus, it establishes that "Top management shall demonstrate leadership and commitment with respect to customer focus by ensuring that the risks and opportunities that can affect conformity of products and services and the ability to enhance customer satisfaction are determined and addressed."

Under Section 6 Planning, Subsection 6.1 Actions to address risk and opportunities, it includes as part of Subsection 6.1.1, "When planning for the quality management system, the organization shall consider the issues referred to in 4.1 and the requirements referred to in 4.2 and determine the risks and opportunities that need to be addressed to:

a. give assurance that the quality management system can achieve its intended result(s);

b. enhance desirable effects;

c. prevent, or reduce, undesired effects;

d. achieve improvement."

Under Subsection 6.1.2, it establishes that the organization shall plan actions to address these risks and opportunities, and how to integrate and implement the actions into its QMS processes and evaluate the effectiveness of these actions. It also establishes that "Actions taken to address risks and opportunities shall be proportionate to the potential impact on the conformity of products and services."

Under Note 1, it establishes that "Options to address risks can include avoiding risk, taking risk in order to pursue an opportunity, eliminating the risk source, changing the likelihood or consequences, sharing the risk, or retaining risk by informed decision."

Under Section 9 Performance evaluation, Subsection 9.1 Monitoring, measurement, analysis and evaluation, Subsection 9.1.3 Analysis and evaluation, it establishes that "The organization shall analyze and evaluate appropriate data and information arising from monitoring and measurement. The results of analysis shall be used to evaluate … the effectiveness of actions taken to address risks and opportunities."

Under Section 9.3 Management review, Subsection 9.3.2 Management review inputs, it establishes that "The management review shall be planned and carried out taking into consideration …the effectiveness of actions taken to address risks and opportunities (see 6.1)."

Under Section 10 Improvement, Subsection 10.2 Nonconformity and corrective action, Subsection 10.2.1, it establishes that "When a nonconformity occurs, including any arising from complaints, the organization shall … update risks and opportunities determined during planning, if necessary."

This international standard also includes Annex A (informative), Clarification of new structure, terminology and concepts. As part of this annex, the following three elements are tied to risk requirements:

A.4 Risk-based thinking

The concept of risk-based thinking has been implicit in previous editions of this International Standard, e.g. through requirements for planning, review, and improvement. This International Standard specifies requirements for the organization to understand its context (see 4.1) and determine risks as a basis for planning (see 6.1). This represents the application of risk-based thinking to planning and implementing quality management system processes (see 4.4) and will assist in determining the extent of documented information.

One of the key purposes of a quality management system is to act as a preventive tool. Consequently, this International Standard does not have a separate clause or subclause on preventive action. The concept of preventive action is expressed through the use of risk-based thinking in formulating quality management system requirements.

The risk-based thinking applied in this International Standard has enabled some reduction in prescriptive requirements and their replacement by performance-based requirements. There is greater flexibility than in ISO 9001:2008 in the requirements for processes, documented information and organizational responsibilities.

Although 6.1 specifies that the organization shall plan actions to address risks, there is no requirement for formal methods for risk management or a documented risk management process. Organizations can decide whether or not to develop a more extensive risk management methodology than is required by this International Standard, e.g. through the application of other guidance or standards.

Not all the processes of a quality management system represent the same level of risk in terms of the organization's ability to meet its objectives, and the effects of uncertainty are not the same for all organizations. Under the requirements of 6.1, the organization is responsible for its application of risk-based thinking and the actions it takes to address risk, including whether or not to retain documented information as evidence of its determination of risks.

A.5 Applicability

This International Standard does not refer to "exclusions" in relation to the applicability of its requirements to the organization's quality management system. However, an organization can review the applicability of requirements due to the size or complexity of the organization, the management model it adopts, the range of the organization's activities and the nature of the risks and opportunities it encounters.

A.8 Control of externally provided processes, products and services

The organization can apply risk-based thinking to determine the type and extent of controls appropriate to particular external providers and externally provided processes, products and services.

Risk Management and Preventive Actions

One of the key purposes of implementing a QMS is to act as a preventive tool. As a result, the formal requirement related to preventive action (existing since the 1994 version of ISO 9001) was removed in the 2015 version and replaced with the requirement for risk-based thinking.

ISO 9001:2015 emphasizes proactive error-proofing processes to prevent problems from occurring because detection (along with root cause identification and corrective action implementation) is not as effective as it should be. Therefore, the aim of the new standard is that planning the QMS up front is the best preventive action, making a clause tied to corrective action unnecessary.

The former corrective action clause (8.5.2 in ISO 9001:2008) has been replaced by a new clause named "Nonconformity and corrective action" (10.2), while the preventive action clause (8.5.3 in ISO 9001:2008) has been deleted and its spirit has been incorporated as part of a new clause (6.1) named "Actions to address risks and opportunities."

The spirit of the new ISO 9001:2015 standard is already present in concepts such as pharmaceutical quality by design and medical device design controls. Risk-based thinking and preventive actions are key parts of the design and development process. Moreover, the strong emphasis on risk management in this new version of the standard has already been embraced by the regulated industry with the incorporation of such concepts through guidances (ICH Q9 for pharmaceuticals) and international standards (ISO 14971 for medical devices or ISO 22000 for food).

Organizations rarely apply the preventive action concept at the optimal stage in a QMS because of cost or time limitations. Preventing every potential problem and nonconformance is prohibitively expensive or even impossible. ISO 9001:2015's requirement of risk-based thinking can help to prevent major failures and issues, but it's unlikely that it will prevent a vast majority of potential problems. I believe that our current understanding of the preventive action concept (when correctly interpreted and applied) will survive for many years. Who can be opposed to extending any identified improvement to other products, processes, or systems not yet affected by an identified root cause(s)?

2

Current Initiatives

FDA-WIDE

FDA oversees the public health standards of an industry that produces trillions of dollars worth of regulated products. The products include the entire food supply (except for meat and poultry), over-the-counter and prescription medications, blood products, vaccines, tissues for transplantation, medical equipment and implantable devices, devices that emit radiation, animal drugs and feeds, and cosmetics. The Agency currently also regulates tobacco products. FDA faces two challenges in assuring that U.S. citizens fully reap the health benefits of modern medical technology and enjoy a safe and nutritious food supply:

1. *FDA must assure the safety and effectiveness of drugs, biologics, food additives, medicated animal feeds, and medical devices before they are allowed on the market.* This challenge means keeping pace with the increasing scientific complexity of products. FDA scientists must accurately judge the readiness of these products to be marketed and help these innovative products reach the market as quickly as possible.

There must also be a sufficient number of reviewers to handle the product review workload in order to provide timely feedback to application sponsors to identify and address deficiencies. Any delays might:

- Postpone critically needed disease prevention and treatment, especially for a growing population of elderly and immune-compromised patients
- Increase the cost of bringing a new product to market by lengthening the time a product is in development
- Result in fewer low-cost alternatives for patients (including children and the elderly)

2. *FDA must also monitor the safety of products once they are on the market.* U.S. consumers use products that are manufactured worldwide and make their way to the market through a wide variety of distribution channels. To ensure that these products are safe, the Agency must oversee their entire life cycle—from production through distribution and use/consumption.

On August 21, 2002, FDA announced a major new initiative pertaining to drug quality regulation: Pharmaceutical cGMPs for the 21st Century: A Risk-Based Approach. In the initial concept paper issued on the initiative, FDA identified "a risk-based orientation" as one of the guiding principles that would drive the initiative. The concept paper stated that "resource limitations prevent uniformly intensive coverage of all pharmaceutical products and production" and that "to provide the most effective public health protection, FDA must match its level of effort against the magnitude of the risk."

The concept paper noted that in the short term, FDA intended to place more emphasis on risk-based allocation of resources for the oversight of drug quality. However, the concept paper also acknowledged that developing more systematic and risk-based approaches would be a long-term effort and that an intermediate step is to "use emerging science and data analysis to enhance compliance programs to target the highest risk areas."

On August 20, 2003, FDA released its strategic action plan, *Protecting and Advancing America's Health*. This strategic plan reflects the Agency's continuing support and priority for the risk management goals of the Pharmaceutical cGMPs for the 21st Century initiative. The first core goal identified in the plan is "efficient risk management" directed toward "ensuring that the Agency's limited resources can provide the most health promotion and protection at the least cost for the public." The discussion of this goal specifically highlights the efforts of the Pharmaceutical cGMPs for the 21st Century initiative as central to the Agency's risk management program. One of the objectives under this goal is to "provide high quality, cost-effective oversight of industry manufacturing, processing, and distribution to reduce risk." Specifically identified under this objective is developing "new inspection approaches to more effectively utilize... resources." Also highlighted is the need to "use emerging science and data analysis to *target the highest risk areas*."

In May 2022, FDA published a draft guidance intended to help stakeholders develop, maintain, and implement risk management plans to proactively assist in the prevention of human drug product and biological product shortages. The following pages describe standards, regulations, and guidance documents that have been published covering risk management activities for the different types of regulated products.

DRUGS AND BIOTECHNOLOGY PRODUCTS

ICH Q9(R1)

In September 2006 FDA adopted the International Conference on Harmonization (ICH) Q9 document entitled "Quality Risk Management," which was revised in May 2023. This guidance provides principles and examples of tools for quality risk management that can be applied to different aspects of pharmaceutical quality. These aspects include development, manufacturing, distribution, inspection, and submission/review processes throughout the life cycle of drug substances, drug products, and biological and biotechnological products (including the use of raw materials, solvents, excipients, packaging and labeling materials in drug products, and biological and biotechnological products).

The manufacturing and use of a drug product, including its components, necessarily entail some degree of risk. The risk to its quality is just one component of the overall risk. Product *quality* must be maintained throughout the *product life cycle* such that the attributes that are important to the quality of the drug product remain consistent with those used in preproduction and clinical studies. An effective quality risk management (QRM) approach can further ensure the high quality of the drug product to the patient by providing a proactive means to identify and control potential quality issues during development and manufacturing. In addition, use of QRM can improve decision-making if a quality problem arises. Effective QRM can facilitate better and more-informed decisions, provide regulators with greater assurance of a company's ability to deal with potential risks, and beneficially affect the extent and level of direct regulatory oversight.

The purpose of this document is to offer a systematic approach to QRM. It serves as a foundation or resource document that is independent of other ICH quality documents, but it supports and complements existing quality practices, requirements, standards, and guidelines within the pharmaceutical industry and regulatory environment.

ICH Q9(R1) provides guidance on the principles and some of the tools of QRM that can enable more effective and consistent risk-based decisions, by both regulators and industry, regarding the quality of drug substances and drug products across the product life cycle. It is not intended to create any new expectations beyond the current regulatory requirements.

It is neither always appropriate nor always necessary to use a formal risk management process (using recognized tools and/or internal procedures, for example, standard operating procedures). The use of informal risk management processes (using empirical tools and/or internal procedures) can also be considered acceptable. Appropriate use of QRM can facilitate but does not obviate industry's obligation to comply with regulatory requirements and does not replace appropriate communications between industry and regulators.

The objectives of the revision were improvement in four areas:

1. Subjectivity in QRM risk assessments and outcomes
2. Insufficient management of supply and product availability risks
3. Lack of understanding of QRM formalities
4. Lack of clarity about risk-based decision-making

According to the ICH business plan, adoption of the revised guideline would result in "more scientific and robust applications of QRM principles, tools, and activities, where subjectivity in QRM outputs is better controlled. This could lead to more science-based manufacturing operations, control strategies, and validation activities, which may result in fewer quality defects and recalls for patients, and potentially in reduced costs for the pharmaceutical industry and healthcare systems."

The revised guideline introduces new text acknowledging that some degree of subjectivity in QRM is unavoidable while offering methods to reduce or control subjectivity. The guideline states that "while subjectivity cannot be completely eliminated from quality risk management activities, it may be controlled by addressing bias, the proper use of quality risk management tools and maximizing the use of relevant data and sources of knowledge."

All personnel involved with QRM activities should "acknowledge, anticipate, and address the potential for subjectivity," according to the revised guidance. Furthermore, decision-makers should ensure that such subjectivity is "controlled and minimized" to allow for more "robust risk-based decision making."

The revision also includes a new section on formality in QRM. The guideline states that "an understanding of formality in quality risk management ... may lead to resources being used more efficiently, where lower risk issues are dealt with via less formal means, freeing up resources for managing higher risk issues and more complex problems that may require increased levels of rigor and effort."

The revision also has new text on the role of QRM in addressing product availability risks. It states that "quality/manufacturing issues, including non-compliance with Good Manufacturing Practice (GMP), are a frequent cause of product availability issues (for example, product shortages). The interests of patients are served by risk-based drug shortage prevention and mitigation activities that help to proactively manage supply chain complexities and ensure availability of needed medicines. An effective pharmaceutical quality system drives both supply chain robustness and sustainable GMP compliance."

Within the revised guideline, the term "risk identification" has been replaced with "hazard identification" to harmonize with discussion in the risk assessment portion of the guidance. The section on risk review, however, remains unchanged.

QbD Guidances

Quality by design (QbD) means designing and developing a product and associated manufacturing processes that will be used during product development to ensure that the product consistently attains a predefined quality at the end of the manufacturing process. It means building quality into the process and product in a systematic, science- and risk-based manner. Quality by design, in conjunction with a quality system, provides a sound framework for the transfer of product knowledge and process understanding from product development to the commercial manufacturing processes and for post-development changes and optimization. This concept is not new. In fact, the 1976 cGMP regulations, when viewed in their entirety, incorporate the concept of quality by design.

The QbD concepts were outlined in the ICH Q8(R2), Q9, and Q10 guidelines. These documents, already adopted by FDA, provide the foundation for implementing QbD during specific stages of product development. Putting the QbD process into place requires several steps. First, you want to define the desired product performance up front and identify product critical quality attributes (CQAs). Next, you design formulation and process to meet product CQAs. A critical component of QbD is understanding the needs of the patient and the specific quality attributes of the product that are linked to safety and efficacy. Thus, to implement QbD it is fundamental to understand the functional relationships between patient needs, product quality attributes, analytical capabilities, and the manufacturing process.

Pharmaceutical companies begin by identifying product requirements for meeting patient needs and then determining the *quality target product profile* (QTPP) and the CQAs required to meet those patient needs.

Based on this information, the company then designs the molecule, the manufacturing process, and the control strategy to ensure the desired product quality is met consistently. The QbD approach can be maintained throughout the life cycle of the product to facilitate innovation and continuous improvement based on expanded knowledge and new technologies. This knowledge is gained during development and grows with more manufacturing experience through process characterization, scale-up, technology transfer, and manufacture, as well as through increased patient exposure to the product. This approach encourages incorporating prior process and product knowledge and experience-based methodologies into manufacturing throughout the life of the product.

When a product is developed using a QbD approach, the impact of raw and starting materials and process parameters on product quality are well understood, and the sources of process and product variability are well known and controlled. In contrast, traditional pharmaceutical manufacturing relies heavily on end product testing, and the process typically lacks the flexibility needed to respond to variables encountered during manufacturing. The outcome is often "quality by conformance" and a rigid, static manufacturing operation that is more susceptible to multiple manufacturing changes post licensure, thus requiring continuous updating of the application through burdensome global regulatory supplements, variations, or amendments.

QbD is a systematic approach to product and process design and development, and it includes the following key elements:

- Targeting the product profile
- Determining the CQAs
- Linking raw material attributes and process parameters to CQAs and performing risk assessment
- Developing a design space
- Designing and implementing a control strategy
- Managing the product life cycle, including continual improvement

Critical Quality Attributes. A CQA is a physical, chemical, biological, or microbiological property or characteristic that should be within an appropriate limit, range, or distribution to ensure the desired product quality. CQAs are generally associated with the drug substance, excipients, intermediates (in-process materials), and drug product.

CQAs of solid oral dosage forms are typically those aspects affecting product purity, strength, drug release, and stability. CQAs for other delivery systems can additionally include more product-specific aspects, such as aerodynamic properties for inhaled products, sterility for parenterals, and adhesion properties for transdermal patches. For drug substances, raw materials, and intermediates, the CQAs can additionally include those properties that affect drug product CQAs (for example, particle size distribution, bulk density).

Potential drug product CQAs derived from the QTPP and/or prior knowledge are used to guide the product and process development. The list of potential CQAs can be modified when the formulation and manufacturing processes are selected and as product knowledge and process understanding increase. Quality risk management can be used to prioritize the list of potential CQAs for subsequent evaluation. Relevant CQAs can be identified by an iterative process of quality risk management and experimentation that assesses the extent to which their variation can have an impact on the quality of the drug product:

- The definition of "critical" varied.
- Most of the applications related CQAs to those attributes that have a direct impact on product quality, safety, and efficacy.
- Most applications included CQAs for the drug product, but only some applications included CQAs for the drug substance.
- Some applications considered properties of the drug substance affecting downstream manufacturing as a CQA.
- Most applications related critical process parameters to CQAs.
- Most applications considered a process parameter noncritical if well controlled or well within the design space.

Risk Assessments. Risk assessment is a valuable science-based process used in quality risk management [see ICH Q9(R1)] that can aid in identifying which material attributes and process parameters potentially have an effect on product CQAs. Risk assessment is typically performed early in the pharmaceutical development process and is repeated as more information becomes available and greater knowledge is obtained.

Risk assessment tools can be used to identify and rank parameters (for example, process, equipment, input materials) with the potential to have an impact on product quality, based on prior knowledge and initial experimental data. The initial list of potential parameters can be quite extensive but can be modified and prioritized by further studies. The list can be refined further through experimentation to determine the significance of individual variables and potential interactions. Once the significant parameters are identified, they can be further analyzed to achieve a higher level of process understanding.

With FDA moving forward with the QbD concept, risk management has become a priority process in the manufacture of pharmaceuticals. QbD, according to ICH Q8(R), is based on sound science and quality risk management. It's a systematic approach to development that begins with predefined objectives and an emphasis on product process understanding and process control. In order to achieve this, a risk management process has to be a priority.

Tools used in the risk assessment include the fishbone diagram, FMEA, and Pareto analysis, among others. A fishbone diagram is used to identify all potential variables, such as raw materials, compression parameters, and environmental factors, that can have an impact on a particular CQA, such as tablet hardness. An FMEA can then be used to rank the variables based on risk (a combination of probability, severity, and detectability) and to select the process parameters with higher risks for further studies to gain a greater understanding of their effects on CQAs. The risk assessment is prepared by a multidisciplinary team based on prior knowledge and experiments.

Through the process of risk assessment and risk control, you want to understand the impact of material attributes and process parameters on product CQAs and then identify and control sources of variability in material and the process.

The use of risk management in QbD will help to identify and link material attributes and process parameters to CQAs, enable early identification of risk to guide developmental efforts, and quantify the severity of harm and likelihood of occurrence. The benefits of risk assessment are numerous. It can be used as a systematic approach to identifying and classifying parameters in the design space—useful in process development, monitoring, improvement, and troubleshooting—and it is used to increase the assurance of quality.

It is important to provide a systematic risk analysis of how raw materials, process steps, and process parameters affect product quality. The risk assessment that leads to the development of a comprehensive control strategy to reduce risk to product quality should be described, and the risk reduction and control should be discussed for changes that occur inside or outside the design space. Risk assessment can provide increased assurance to quality, because when process variability is identified and its linkage to product CQAs is understood, process and product controls reduce the impact of variability, and quality product will continue to be made when movement within the design space occurs in the future.

A risk assessment also is important for effective communication between FDA and industry and for intracompany communication (such as between research/development and manufacturing and between multiple manufacturing sites), and within FDA, risk assessment allows for a dialogue between pre- and post-marketing review functions and between review, compliance, and field inspection staffs.

Design Space. FDA defines *design space* in its Guidance for Industry Q8 (2): Pharmaceutical Development (Food and Drug Administration 2009b) as "The multidimensional combination and interaction of input variables (e.g., material attributes) and process parameters that have been demonstrated to provide assurance of quality. Working within the design space is not considered as a change. Movement out of the design space is considered to be a change and would normally initiate a regulatory post-approval change process. Design space is proposed by the applicant and is subject to regulatory assessment and approval."

The idea behind design space is that process variability leads to product variability. Monitoring the parameters of drug attributes results in less process variability and, ultimately, less product variability. The advantage of the design space is that it provides a wider range within which material attributes can vary in the future without supplemental regulatory approval.

In establishing and presenting a design space, some key points to consider are:

- The effect of formulation-component properties on process performance and product quality
- Multivariate interactions
- Scale-up and equipment issues
- The effect of operation or site change
- Uncertainty addressed with risk management

A control strategy is designed to ensure that a product of required quality will be produced consistently. The elements of the control strategy should describe and justify how in-process controls and the controls on input materials (drug substances and excipients), intermediates (in-process materials), container closure systems, and drug products contribute to the final product quality. These controls should be based on product, formulation, and process understanding and should include, at a minimum, control of the *critical process parameters* and material attributes.

A control strategy can include, but is not limited to, the following:

- Control of input material attributes (for example, drug substances, excipients, primary packaging materials) based on an understanding of their impact on manufacturability or product quality
- Product specifications
- Controls for unit operations that have an impact on downstream processing or product quality (for example, the impact of drying on degradation, particle size distribution of the granulate on dissolution)
- In-process or real-time release testing in lieu of end product testing (measurement and control of CQAs during processing)
- A monitoring program (full product testing at regular intervals) for verifying multivariate prediction models

A control strategy can include different elements. For example, one element of the control strategy could rely on end product testing whereas another could depend on real-time release testing. The rationale for using these alternative approaches should be described in the submission.

Other points that can be considered include:

- Better use of process knowledge and understanding in setting specifications
- More meaningful sampling for drug product testing
- Experience needed in setting real-time release specifications

REMS and Medication Guides

The Food and Drug Administration Amendments Act of 2007 (FDAAA) created new section 505-1 of the Federal Food, Drug, and Cosmetic Act (FD&C Act) (21 U.S.C. 355-1), which authorizes FDA to require a Risk Evaluation and Mitigation Strategy (REMS) when necessary to ensure that the benefits of a drug outweigh the risks. Under section 505-1(e), FDA may require that a REMS for a drug include one or more of the elements described in the subsection, including when the criteria in 21 CFR part 208 are met—the requirement for an applicant to develop a Medication Guide for distribution to each patient when the drug is dispensed.

Under part 208, Medication Guides may be safety-related, addressing serious risk(s) (relative to benefits) of which patients should be made aware, and/or efficacy-related, when patient adherence to directions for use is crucial to the drug's effectiveness. Since the enactment of FDAAA, FDA has considered any new Medication Guide (or safety-related changes to an existing Medication Guide) to be part of a REMS. However, the Agency has the authority to determine, based on the risks of a drug and public health concern, whether a Medication Guide should be required as part of a REMS when the standard in part 208 is met, and may decide that the Medication Guide should be required as labeling but not part of a REMS if FDA determines that a REMS is not necessary to ensure that the benefits of the drug outweigh its risks.

Section 208.1(a) states that Medication Guides apply primarily to human prescription drug products used on an outpatient basis without direct supervision by a healthcare professional, and are applicable to both new and refill prescriptions. Section 208.1(c) authorizes FDA to require a Medication Guide if FDA determines that one or more of the following circumstances exist:

1. The drug product is one for which patient labeling could help prevent serious adverse effects.
2. The drug product is one that has serious risk(s) (relative to benefits) of which patients should be made aware because information concerning the risks could affect the patients' decision to use, or continue to use, the product.
3. The drug product is important to health, and patient adherence to directions for use is crucial to the drug's effectiveness.

Part 208 specifies the content and format of Medication Guides and manufacturer requirements to provide Medication Guides for distribution. Manufacturers of drug products for which a Medication Guide is required must both:

- Obtain FDA approval of the Medication Guide before the Medication Guide is provided
- Ensure that Medication Guides are provided in sufficient numbers, or provide the means to produce Medication Guides in sufficient numbers, to distributors, packers, and authorized dispensers to permit the authorized dispenser to provide a Medication Guide to each patient receiving a prescription for the drug product

Part 208 also specifies the requirements for distribution of Medication Guides:

- Distributors and packers who receive the Medication Guides, or the means to produce Medication Guides, must provide the Medication Guides or the means to provide them to authorized dispensers.

- Each authorized dispenser of a prescription drug product for which a Medication Guide is required must provide the Medication Guide directly to each patient or each patient's agent when the product is dispensed, unless an exemption applies.

Sections 208.3(a) and (b) contain the following definitions:

- *Authorized dispenser* is "an individual licensed, registered, or otherwise permitted by the jurisdiction in which the individual practices to provide drug products on prescription in the course of professional practice."
- *Dispense to patients* means the "act of delivering a prescription drug to a patient or an agent of the patient either:
  - (1) By a licensed practitioner or an agent of a licensed practitioner, either directly or indirectly, for self-administration by the patient, or the patient's agent, or outside the licensed practitioner's direct supervision; or
  - (2) By an authorized dispenser or an agent of an authorized dispenser under a lawful prescription of a licensed practitioner."

Between March 25, 2008, when the REMS provisions of FDAAA took effect, and January 1, 2011, FDA approved more than 150 Medication Guides for products approved under new drug applications (NDAs) and biologic license applications (BLAs) as part of a REMS. One hundred and eight of these REMSs included only a Medication Guide and a timetable for submission of assessments of the REMS. In some cases, Medication Guides have been approved as part of a REMS for drugs that are often provided in inpatient settings or in outpatient settings where the drug is dispensed to a healthcare professional who then administers the drug to the patient. Questions have arisen concerning FDA's policy on whether the Medication Guide must be provided every time the drug is dispensed because part 208 states that the regulations are intended to apply primarily in the outpatient setting. As of December 31, 2023, there are 67 currently approved individual and shared system REMS.[1]

Risk Management Plans to Mitigate the Potential for Drug Shortages

In May 2022, FDA published a draft guidance for industry named "Risk Management Plans to Mitigate the Potential for Drug Shortages." This guidance is intended to help stakeholders develop, maintain, and implement risk management plans (RMPs) to proactively assist in the prevention of human drug and biological products shortages. RMPs can provide stakeholders with a framework to proactively identify, prioritize, and implement strategies to mitigate hazards that can cause a supply disruption. Such a supply disruption may lead to a drug shortage. Effective quality risk management can facilitate better, more informed decisions; provide FDA with greater assurance that stakeholders understand and can manage the associated risks; and potentially affect the extent and level of direct regulatory oversight. Based on recent publications and reports, the majority of drug shortages are associated with quality issues. This guidance describes a framework for stakeholders to consider when developing RMPs that aligns with principles stated in the International Council for Harmonization (ICH) guidance for industry Q9 Quality Risk Management (first adopted 2006). In addition, FDA also recommends risk factors to consider when developing the content of the RMPs.

FDA recognizes that drug shortages pose a significant public health threat because they can delay, and in some cases even deny, critically needed care for patients. Proactive risk assessment can enable stakeholders to predict and prevent supply disruptions that could potentially lead to a drug shortage.

In July 2012, with the enactment of the Food and Drug Administration Safety and Innovation Act (FDASIA), Congress empowered FDA with tools to work in collaboration with industry to help prevent or mitigate drug supply disruptions and drug shortages, and clarified cGMP requirements relevant to the oversight and controls over the manufacture of drugs to ensure quality. In 2011 and 2012, FDA assisted in mitigating 368 new drug shortages and preventing 473 drug shortages. In contrast, in the six-year span after FDASIA was enacted, FDA assisted in mitigating 222 new drug shortages and preventing 806 drug shortages. Although the number of new drug shortages has declined significantly since the peak in 2011, reaching a low in 2015 and 2016, this downward trend did not continue in subsequent years. Drug shortages continue to occur and at roughly the same levels since 2018.

Further, drug shortages have grown more persistent (for example, increased length of active drug shortages). Many of the reasons for drug shortages are, for example, issues related to drug quality, disruptions to supply chain manufacturing operations (for example, caused by natural disaster or discontinuation of components by suppliers), limitations in forecasting future demand, and market withdrawals of drug products. Additionally, in recent years, FDA has observed a rise in the number of cyberattacks on drug manufacturers and is increasingly concerned about the effect of such attacks on the drug supply chain.

In March 2020, with the enactment of the Coronavirus Aid, Relief, and Economic Security Act (CARES Act), U.S. Congress added section 506C(j) to the Federal Food, Drug, and Cosmetic Act (FD&C Act), which requires certain manufacturers to develop, maintain, and implement, as appropriate, a "redundancy risk management plan that identifies and evaluates risks to the supply of the drug, as applicable, for each establishment in which such drug or active pharmaceutical ingredient of such drug is manufactured." Section 506C(j) became effective on September 23, 2020.

MEDICAL DEVICES

Risk management is an integral part of any medical device's life cycle because it details the hazards and risks linked to the use of the device. It also describes how manufacturers handled those hazards and risks to optimize its safety and performance. Typically, manufacturers created the risk management file of each device by continuously analyzing the risk of the device throughout the design and development processes.

One of the most significant factors associated with safe and effective medical devices is the identification of risks and the application of risk management tools to control them. To ensure that medical devices are safe and effective, risk management is a regulatory requirement worldwide. U.S. FDA mandates it in the Quality System Regulation (21 CFR Part 820). Europe requires it in the Medical Device Regulation (2017/745) and the In Vitro Diagnostic Devices Regulation (2017/746). Canada Medical Devices Regulations (SOR/98-282) categorizes devices into four classes based on the risk. All countries have some requirement for risk management referenced in their national regulations and/or ISO 13485:2016.

Current versions of two important medical devices' international standards are discussed below: ISO 13485:2016 and ISO 14971:2019.

ISO 13485:2016 Risk Requirements

ISO 13485, *Medical devices—Quality management systems—Requirements for regulatory purposes,* is an international standard that specifies the requirements for a quality management system specific to the medical devices industry. It was revised in March 2016. The new version has a greater emphasis on risk management and risk-based decision-making, as well as changes related to the increased regulatory requirements for organizations in the supply chain.

There are many elements of risk-based thinking in the ISO 13485:2016 international standard that may affect organizations as they work toward compliance to the standard. The following excerpts and summaries describe references to risk in this international standard.

Under Section 0.2, Clarification of concepts, the standard establishes that when a requirement is qualified by the phrase "as appropriate," it is deemed to be appropriate unless the organization can justify otherwise. A requirement is considered appropriate if it is necessary for the organization to manage risks. It also states that "when the term 'risk' is used, the application of the term within the scope of this International Standard pertains to safety or performance requirements of the medical device or meeting applicable regulatory requirements."

Under Section 3, Terms and Definitions, it describes *risk* as a "combination of the probability of occurrence of harm and the severity of that harm" and *risk management* as "systematic application of management policies, procedures and practices to the tasks of analyzing, evaluating, controlling and monitoring risk."

Under Section 4, Quality management system, Subsection 4.1, General requirements, Subsection 4.1.2, "The organization shall apply a risk-based approach to the control of the appropriate processes needed for the quality management system," and under Subsection 4.1.5, "When the organization chooses to outsource any process that affects product conformity to requirements, it shall monitor and ensure control over such processes. The organization shall retain responsibility of conformity to this International Standard and to customer and applicable regulatory requirements for outsourced processes. The controls shall be proportionate to the risk involved and the ability of the external party to meet the requirements in accordance with 7.4. The controls shall include written quality agreements."

Under Subsection 4.1.6, it establishes that "The specific approach and activities associated with software validation and revalidation shall be proportionate to the risk associated with the use of the software."

Under Section 6, Resource management, Subsection 6.2, Human resources, a Note establishes that "The methodology used to check effectiveness is proportionate to the risk associated with the work for which the training or other action is being provided."

Under Section 7, Product realization, Subsection 7.1, Planning of product realization, it establishes that "The organization shall document one or more processes for risk management in product realization. Records of risk management activities shall be maintained."

Under Subsection 7.3.3, Design and development inputs, it establishes that "Inputs relating to product requirements shall be determined and records maintained (see 4.2.5). These inputs shall include applicable output(s) of risk management."

Under Subsection 7.3.9, Control of design and development changes, it establishes that "The review of design and development changes shall include evaluation of the effect of the changes on constituent parts and product in process or already delivered, inputs or outputs of risk management and product realization processes."

Under Section 7.4, Purchasing, Subsection 7.4.1, Purchasing process, it establishes that "The organization shall establish criteria for the evaluation and selection of suppliers. The criteria shall be proportionate to the risk associated with the medical device," and it also states that "Non-fulfilment of purchasing requirements shall be addressed with the supplier proportionate to the risk associated with the purchased product and compliance with applicable regulatory requirements."

Under Subsection 7.4.3, Verification of purchased product, it establishes that "The organization shall establish and implement the inspection or other activities necessary for ensuring that purchased product meets specified purchasing requirements. The extent of verification activities shall be based on the supplier evaluation results and proportionate to the risks associated with the purchased product."

Under Subsection 7.5.6, Validation of processes for production and service provision, it establishes that "The specific approach and activities associated with software validation and revalidation shall be proportionate to the risk associated with the use of the software, including the effect on the ability of the product to conform to specifications."

Under Section 7.6, Control of monitoring and measuring equipment, it establishes that "The specific approach and activities associated with software validation and revalidation shall be proportionate to the risk associated with the use of the software, including the effect on the ability of the product to conform to specifications.

Under Section 8.2, Monitoring and measurement, Subsection 8.2.1, Feedback, it establishes that "The information gathered in the feedback process shall serve as potential input into risk management for monitoring and maintaining the product requirements as well as the product realization or improvement processes."

ISO 14971:2019

ISO 14971 is an international risk management standard for medical devices (including in vitro diagnostic medical devices). It defines a set of medical device risk management requirements. The purpose of ISO 14971 is to help manufacturers establish a medical device risk management process that can be used to identify hazards, estimate and evaluate risks, and implement and monitor the effectiveness of risk control measures (ISO 2019a).

ISO 14971 was first published in 2000. The most recent edition is the third edition, which was published in December of 2019. ISO 14971 was developed because no standard existed at the time, and regulators throughout the world wanted manufacturers to apply risk management to their medical devices.

This third edition cancels and replaces the second edition (ISO 14971:2007), which was technically revised. The main changes compared to the previous edition are as follows:

- A clause on normative references is included, it is copied from the Foreword.
- The defined terms are updated, and many are derived from ISO/IEC Guide 63:2019. Defined terms are printed in italics to assist the reader in identifying them in the body of the document.
- Definitions of benefit, reasonably foreseeable misuse, and state of the art are introduced.
- More attention is given to the benefits that are expected from the use of the medical device. The term benefit-risk analysis is aligned with the terminology used in some regulations.
- It is explained that the process described in ISO 14971 can be used for managing risks associated with medical devices, including those related to data and systems security.
- The method for evaluating the overall residual risk and the criteria for its acceptability are required to be defined in the risk management plan. The method can include gathering and reviewing data and literature for the medical device and for similar medical devices and other similar products on the market. The criteria for the acceptability of the overall residual risk can be different from the criteria for acceptability of individual risks.
- The requirements to disclose residual risks were moved and merged into one requirement, after the overall residual risk is evaluated and judged acceptable.
- The review before commercial distribution of the medical device concerns the execution of the risk management plan. The results of the review are documented as the risk management report.

– The requirements for production and post-production activities were clarified and restructured. More detail is given on the information to be collected and the actions to be taken when the collected information has been reviewed and determined to be relevant to safety.

– Several informative annexes were moved to the guidance in ISO/TR 24971:2020, which was revised in parallel. More information and a rationale for the requirements in this third edition of ISO 14971 are provided in Annex A. The correspondence between the clauses of the second edition and those of this third edition are given in Annex B.

ISO 14971:2019 specifies a process for a manufacturer to identify the hazards associated with medical devices, estimate and evaluate the associated risks, control these risks, and monitor the effectiveness of the controls. It applies to product design, production, and all post-production activities. Since risks can be introduced throughout the product life cycle and risks that emerge at one point can often be controlled at a completely different point, ISO 14971 must be used throughout the product's existence.

However, ISO 14971 does not apply to clinical decision-making. Whether a medical device should be used in the context of a particular clinical procedure is a matter of judgment. Such judgments are usually made by qualified healthcare professionals and are not regulated by this international standard. The standard requires that top management demonstrate its commitment to risk management by:

- Establishing the process
- Setting the policy for acceptable risk
- Reviewing the effectiveness of the process at determined intervals

Medical device risk management is mandatory in most countries and regions of the world. Every regulated medical device manufacturer needs to be able to prove to regulators that it uses risk management methods to control risk and ensure that its products are safe. For example, ISO 14971 is widely recognized as the official standard for medical device risk management. Regulators in most major markets expect medical device manufacturers to use this standard to manage risk. ISO 14971 has been officially recognized by the U.S. FDA and by Health Canada. In addition, the European Union has adopted it as a harmonized standard, and Australia and Japan have accepted it as the official standard for medical device risk management. ISO 14971 does not expect medical device manufacturers to become certified. Manufacturers can simply use the standard to manage risk and then declare that they are in compliance (if they can demonstrate that this is in fact true).

ISO 14971:2019 describes a risk management process (Parts 4 through 10) and is organized in the following way:

Part 4. Establish a risk management framework.

Part 5. Perform a risk analysis for each medical device.

Part 6. Evaluate risk for each identified hazardous situation.

Part 7. Develop risk control measures when risk must be reduced.

Part 8. Evaluate the overall residual risk posed by each device.

Part 9. Review the risk management process and prepare a report.

Part 10. Monitor the device during production and post-production.

ISO 14971:2019 does not change the risk management process, but it finally defines the "benefits," which the previous versions did not do. *Benefit* is now defined as: "Positive impact or desirable outcome of the use of a medical device on the health of an individual, or a positive impact on patient management or public health."

ISO/TR 24971:2020. This technical report will assist manufacturers in the development, implementation, and maintenance of a risk management process for medical devices that aims to meet the requirements of ISO 14971:2019. It provides guidance on the application of ISO 14971:2019 for a wide variety of medical devices. These medical devices include active, nonactive, implantable, and nonimplantable medical devices; software as medical devices; and in vitro diagnostic medical devices.

The clauses and subclauses in this document have the same structure and numbering as the clauses and subclauses of ISO 14971:2019 to facilitate the use of this guidance in applying the requirements of the standard. Further division into subclauses is applied where considered useful. The informative annexes contain additional guidance on specific aspects of risk management.

Annex H was prepared in cooperation with Technical Committee ISO/TC 212, Clinical laboratory testing and in vitro diagnostic test systems.

This document describes approaches that manufacturers can use to develop, implement, and maintain a risk management process conforming to ISO 14971:2019. Alternative approaches can also satisfy the requirements of ISO 14971:2019.

ISO 14971 and FDA. FDA recognized ISO 14971:2019 as an acceptable risk management model under Recognition Number 5-125 dated December 23, 2019.

GHTF/IMDRF Guidelines

The Global Harmonization Task Force (GHTF) was a voluntary group of representatives from national medical device regulatory authorities (such as the U.S. FDA) and the members of the medical device industry whose goal was the standardization of medical device regulation across the world. The representatives from its five founding members (the European Union, the United States, Canada, Japan, and Australia) were divided into three geographical areas: Europe, Asia-Pacific, and North America. Each of these actively regulates medical devices using its own unique regulatory framework. Founded in 1992, the GHTF was created by EU, U.S., Japanese, Australian, and Canadian regulatory and industry officials to respond to the growing need for international harmonization in the regulation of medical devices. In November 2012, the global medical device harmonization body ceased its activities, leaving its unfinished work to the International Medical Device Regulators Forum (IMDRF), a successor organization comprising officials from regulatory agencies—not industry—around the world. The organization had been a mainstay within the regulatory harmonization movement.

GHTF guideline SG3/N15R8 "Implementation of Risk Management Principles and Activities within a Quality Management System" was released in May 2005 (http://www.imdrf.org). This document is intended to assist medical device manufacturers with the integration of a risk management system or risk management principles and activities into their existing QMS by providing practical explanations and examples. Medical device manufacturers are generally required to have a QMS as well as processes for addressing device-related risks. These processes for managing risk can evolve into a stand-alone management system. While manufacturers may choose to maintain these two management systems separately, it may be advantageous to integrate them, as it could reduce costs, eliminate redundancies, and lead to a more effective management system.

A basic understanding of QMS requirements and a basic knowledge of quality management system terminology are assumed in this guidance document. This document is based on general principles of a QMS and general principles of a risk management system. This document also:

- Has general applicability to QMSs for manufacturers providing medical devices
- Discusses risk management related to medical device safety, rather than financial or other business risks
- Does not suggest a particular method of implementation

- Does not include requirements to be used as the basis of regulatory inspection or certification assessment activities

The scope of the medical device manufacturer's QMS will define the applicability and extent of implementing risk management principles and activities. Processes required by the QMS and performed by suppliers to the manufacturer are the responsibility of the manufacturer. Risk management activities relating to any process within the QMS are ultimately the responsibility of the manufacturer.

An effective QMS is essential for ensuring the safety and performance of medical devices. A well-defined QMS includes safety considerations in specific areas. Given the importance of safety, it is useful to identify some key activities that specifically address safety issues and ensure appropriate input and feedback from these activities into the QMS. The degree to which safety considerations are addressed should be commensurate with the degree of the risk and the nature of the device.

Risk management principles should be applied throughout the life cycle of medical devices and used to identify and address safety issues. In general, risk management can be characterized by phases of activities. The following discussion is one example of how these phases can be described:

a. The *first* phase can be the determination of levels of risk that would be acceptable in the device. Manufacturers should have a procedure or policy to determine risk acceptability criteria. These *risk acceptability* criteria may come from an analysis of the manufacturer's own experience with similar medical devices or research on what appear to be currently accepted risk levels by regulators, users, or patients, given the benefits derived from diagnosis or treatment with the device. Risk acceptability criteria generally should be reflective of the state of the art in controlling risks.

b. The *second* phase can be *risk analysis*. This phase starts with identifying hazards that may occur due to characteristics or properties of the device during normal use or foreseeable misuse. After hazards are identified, risks are estimated for each of the identified hazards, using available information.

c. In the *third* phase, the estimated risks are compared to the risk acceptability criteria. This comparison will determine an appropriate level of risk reduction, if necessary. This is called *risk evaluation*. The combination of risk analysis and risk evaluation is called *risk assessment*.

d. The *fourth* phase can be composed of *risk control* and monitoring activities. The manufacturer establishes actions (risk control measures) intended to eliminate or reduce each risk to meet the previously determined risk acceptability criteria. Within the limits of feasibility, one or more risk control measures may be incorporated in order to achieve this end. Risk control activities may begin as early as design input and continue through the design and development process, manufacturing, distribution, installation, and servicing, and throughout the medical device life cycle.

Some regulatory schemes prescribe a fixed hierarchy of risk control measures that should be examined in the following order:

1. Inherent safety by design
2. Protective measures in the device or its manufacture
3. Information for safety, such as warnings, and so on

Throughout the life cycle of the device, the manufacturer monitors whether the risks continue to remain acceptable and whether any new hazards or risks are discovered. Information typically obtained from the QMS—for example, production, complaints, customer feedback—should be used as part of this monitoring. If, at any time, a risk is determined to be unacceptable, the existing risk analysis should be reexamined and appropriate action should be taken to meet the risk acceptability criteria. If a new hazard is identified, the four phases of risk management should be performed.

FDA has recognized three classes of medical devices based on the level of control necessary to assure the safety and effectiveness of the device (see Table 2.1). The classification procedures are described in the Code of Federal Regulations, Title 21, part 860 (usually known as 21 CFR 860).

Non–U.S. Regulations

As a direct example of how some countries apply risk management criteria to regulate medical products, I can mention the Canadian *Guidance on Risk Classification of Medical Device Observations* (GUI-0079) implemented October 1, 2010 (currently version 3, issued December 2022). The purpose of this guidance is to classify the observations noted during inspections at medical device establishments under the Medical Devices Inspection Program according to their risk (Health Canada 2022).

Table 2.1 FDA medical device classification.

Class I **General controls**	**Class II** **General controls with special controls**	**Class III** **General controls and pre-market approval**
Class I devices are subject to the least regulatory control. General controls include provisions that relate to adulteration; misbranding; device registration and listing; pre-market notification; banned devices; notification, including repair, replacement, or refund; records and reports; restricted devices; and current good manufacturing practices (cGMP). Class I devices are not intended for use in supporting or sustaining life or to be of substantial importance in preventing impairment to human health, and they may not present a potential unreasonable risk of illness or injury. Most Class I devices are exempt from the pre-market notification and/or good manufacturing practices regulation.	Class II devices are those for which general controls alone are insufficient to assure safety and effectiveness, and existing methods are available to provide such assurances. In addition to complying with general controls, Class II devices are also subject to special controls. A few Class II devices are exempt from the pre-market notification. Special controls may include special labeling requirements, mandatory performance standards, and post-market surveillance. Devices in Class II are held to a higher level of assurance than Class I devices, and are designed to perform as indicated without causing injury or harm to patient or user.	A Class III device is one for which insufficient information exists to assure safety and effectiveness solely through the general or special controls sufficient for Class I or Class II devices. Such a device needs pre-market approval, a scientific review to ensure the device's safety and effectiveness, in addition to the general controls of Class I. Class III devices are usually those that support or sustain human life, are of substantial importance in preventing impairment of human health, or that present a potential, unreasonable risk of illness or injury.
Examples: elastic bandages, examination gloves, and handheld surgical instruments	Examples: powered wheelchairs, infusion pumps, and surgical drapes	Examples: implantable pacemaker, pulse generators, HIV diagnostic tests, and automated external defibrillators

Its scope is to ensure uniformity among the inspectors in the attribution of the overall establishment rating following establishment inspections, and to inform the industry of situations that the inspectorate considers unacceptable and that will generate a noncompliant rating following an inspection.

This guidance explains the risk classification assigned to observations made during inspections of medical device manufacturers, importers, and/or distributors. The risk classification of an observation is determined by the inspector through the use of this document and other relevant documents and information, and considers the compliance history of the establishment and the real and/or potential risk involved.

Inspection Ratings and Observations Classifications

Table 2.2 describes the possible outcomes of inspections under Canadian regulation.

Table 2.2 Canadian medical device inspection classification.

Compliant	Noncompliant
At the time of the inspection, the regulated party has demonstrated that the activities it conducts are in compliance with the *Food and Drugs Act* and its associated Regulations. A compliant rating does not mean that there are no observations or corrective actions required.	At the time of the inspection, the regulated party has not demonstrated that the activities it conducts are in compliance with the *Food and Drugs Act* and its associated Regulations.

Observations are classified as risk 1, 2, or 3, where risk 1 represents the greatest risk based on criteria such as health risks and deviation from regulations (see Table 2.3).

Another area where risk management is widely used is the classification of medical product: prescription or nonprescription status for drugs, or how regulatory authorities recognize different classes of medical devices based on their design complexity, their use characteristics, and their potential for harm if misused. Each country or region defines these categories in different ways. The authorities also recognize that some devices are provided in combination with drugs, and regulation of these combination products takes this factor into consideration.

Table 2.3 Canadian medical device risk classification.

Risk 1	Risk 2	Risk 3
An observation describing a situation that has resulted, or has the potential to result in an immediate or latent health risk, and may include any observation that involves fraud, misrepresentation, or falsification of medical devices or records	An observation related to the failure of an establishment to meet the requirements of the Act and the Regulations related to the processing, importation, and/or distribution of medical devices, or a situation in which the probability of an immediate health risk is remote because procedures are in place to mitigate the risk	An observation that is neither Risk 1 nor Risk 2, but is related to a deviation from the Act or the Regulations

Classification of Medical Devices in Canada and the European Union. The Medical Devices Bureau of Health Canada has recognized four classes of medical devices based on the level of control necessary to assure the safety and effectiveness of the device. Class I devices present the lowest potential risk and do not require a license. Class II devices require the manufacturer's declaration of device safety and effectiveness, whereas Class III and IV devices present a greater potential risk and are subject to in-depth scrutiny.

Canadian classes of medical devices generally correspond to the Medical Devices Regulation (EU) 2017/745 devices as follows: Class IV (Canada) generally corresponds to Class III (EU), Class III (Canada) generally corresponds to Class IIb (EU), Class II (Canada) generally corresponds to Class IIa (EU), and Class I (Canada) generally corresponds to Class I (EU). Examples are surgical instruments (Class I); contact lenses, ultrasound scanners (Class II); orthopedic implants, hemodialysis machines (Class III); and cardiac pacemakers (Class IV).

The EU has four classes of medical devices that generally correspond to Canada's four classes, as illustrated in Table 2.4. The classification of medical devices in the European Union is outlined in Medical Devices Regulation (2017/745). There are basically four classes, ranging from low risk to high risk: Class I, Class IIa, Class IIb, and Class III. The European IVD regulation (2017/746) has a similar classification system, from Class A (the lowest risk) to Class D (the highest risk).

Table 2.4 Comparison of Canadian and European medical device classification.

Canadian Medical Devices Regulations		Medical Devices Regulation (EU) 2017/745
Class IV	generally corresponds to	Class III
Class III	generally corresponds to	Class IIb
Class II	generally corresponds to	Class IIa
Class I	generally corresponds to	Class I

The authorization of medical devices in the European Union is guaranteed by a EU MDR Declaration of Conformity. This declaration is issued by the manufacturer itself; medical devices that pertain to Class I (on condition they do not need to be sterilized or are not used to measure a function) can be put on the market purely by self-certification.[2] All others must be verified by a Certificate of Conformity issued by a Notified Body.

The European classification depends on rules that involve the medical device's duration of body contact, its invasive character, its use of an energy source, its effect on the central circulation or nervous system, its diagnostic impact, or its incorporation of a medicinal product.

FOOD SAFETY

Food safety is a scientific discipline describing handling, preparation, and storage of food in ways that prevent foodborne illness. This includes several activities that should be followed to avoid potentially severe health hazards. Foodborne diseases take a major toll on health. Millions of people fall ill and many die because of eating unsafe food. Deeply concerned by this, World Health Organization (WHO) member states adopted a resolution in 2000 to recognize food safety as an essential public health function.

Food safety encompasses actions aimed at ensuring that all food is as safe as possible. Food safety policies and actions need to cover the entire food chain, from production to consumption. Unsafe food poses global health threats, endangering everyone, especially those particularly vulnerable (the elderly, infants, pregnant woman, and so on). For example, every year 220 million children contract diarrheal diseases and 96,000 die (World Health Organization 2015).

In 2002, FDA's Center for Food Safety and Applied Nutrition (CFSAN) recommended the following:

> Risk analysis is a powerful tool that should be used to enhance the scientific basis of regulatory decisions. It should be conducted within CFSAN through the efforts of risk assessment, risk man-

agement, and risk communication teams. Risk assessment should be conducted in an iterative manner that allows refinement of the risk assessment question(s), key assumptions, and data used in the model. The exchange of information (communication) within and between the risk analysis teams, with other agencies, and stakeholders (including industry, consumer groups, and other interested parties) should be encouraged by active participation in the process and collaboration, when appropriate.

To support and promote the use of a risk analysis framework for initiating and conducting 'major' risk assessments, CFSAN should:

1. Adopt a decision-based approach to identify and select risk assessments conducted by CFSAN, particularly those that are 'major' (complex and impact or involve multiple offices). Available resources, regulatory needs, and public health concerns should be considered in the selection of risk assessments. This approach should be implemented for microbial risk assessments now, and later expanded to include chemical and other non-microbial hazards.

2. Establish a procedure for the conduct of risk assessment within a risk analysis framework. The procedure should identify the boundaries and responsibilities of key participants in the risk analysis process.

3. Develop criteria to evaluate the quality of data used for risk assessments and specify what information is needed to scientifically evaluate the usefulness of a study or data set used for risk assessment.

4. Develop guidelines to evaluate risk assessments and supporting data developed by stakeholders and submitted to the Center.

5. Formalize a peer review process that will encourage critical review and evaluation of CFSAN's risk assessments by government and non-government experts in a manner that improves the science and acceptance of complex risk assessments.

6. Build capacity to conduct complex risk assessments by providing training opportunities for current staff, hiring new staff or using contractors (as needed), and acquiring additional resources such as computers, software, and dedicated workspace.

ISO 22000:2018 and Hazard Analysis and Critical Control Points (HACCP)

The launch on September 1, 2005, of the ISO 22000 (Food safety management) series, developed by ISO technical committee ISO/TC 34, Food Products, signaled the arrival of a truly global option for ensuring safe food supply chains. ISO 22000:2018 (second edition issued June 2018) is an international standard that defines the requirements of a food safety management system covering all organizations in the food chain from "farm to fork." ISO 22000:2018 specifies requirements for a food safety management system where an organization in the food chain (including interrelated organizations such as producers of equipment, packaging materials, cleaning agents, additives, and ingredients) needs to demonstrate its ability to control food safety hazards to ensure that food is safe at the time of human consumption (ISO 2018a). The standard combines generally recognized key elements to ensure food safety along the food chain, including:

- Interactive communication
- System management
- Control of food safety hazards through prerequisite programs and HACCP plans
- Continual improvement and updating of the food safety management system

The most effective food safety systems are established, operated, and updated within the framework of a structured management system and incorporated into the overall management activities of the organization. This provides maximum benefit for the organization and interested parties. ISO 22000 has been aligned with ISO 9001 in order to enhance the compatibility of the two standards. ISO developed additional standards that are related to ISO 22000. These standards are known as the ISO 22000 family of standards. Figure 2.1 depicts the current composition of this family of standards, each focusing on different aspects of food safety management.

ISO 22000:2018 major changes include modifications to its structure and clarification of key concepts such as:

- *The high-level structure*. To make life easier for businesses using more than one management system standard, the new version of ISO 22000 now follows the same structure as all the other ISO management system standards, the high-level structure (HLS).

- ISO 22000:2018 contains the overall guidelines for food safety management.
- ISO 22005:2007 focuses on traceability in the feed and food chain.
- ISO/TS 22002-1:2009 contains specific prerequisites for food manufacturing.
- ISO/TS 22002-2:2013 contains specific prerequisites for catering.
- ISO/TS 22002-3:2011 contains specific prerequisites for farming.
- ISO/TS 22002-4:2013 contains specific prerequisites for food packaging manufacturing.
- ISO/TS 22002-5:2019 contains specific prerequisites for food transport and storage.
- ISO/TS 22002-6:2016 contains specific prerequisites for feed and animal food production.
- ISO/TS 22003:2022 provides guidelines for audit and certification bodies.

Figure 2.1 ISO 22000 family of standards.

- *The risk approach*. The standard now includes a different approach to understanding risk.
- *The PDCA cycle*. The standard clarifies the plan–do–check–act (PDCA) cycle by having two separate cycles in the standard working together: one covering the management system and the other covering the principles of HACCP.
- *The operation process*. A clear description is given of the differences between key terms such as: critical control points (CCPs), operational prerequisite programs (OPRPs), and prerequisite programs (PRPs).

Additional new concepts were introduced in ISO 22000:2018:

- The concept of the process approach was made explicit.
- The PDCA cycle is now introduced at two levels: organizational and operational.
- The scope was widened to include food for animals (food is intended for consumption by humans and animals and includes feed and animal food; feed is intended to be fed to food-producing animals; animal food is intended to be fed to non-food-producing animals, such as pets).
- The term "product" was defined and clarified to include service.
- New terms were defined, such as "acceptable level," "action criterion," and "significant food safety hazard."

- Some definitions were revised and improved, for example "control measure," "critical control point (CCP)," "operational prerequisite program (OPRP)," and "prerequisite program (PRP)."
- The understanding of interested parties needs and expectations was introduced.
- The leadership role of the top management (commitment, responsibilities, and authorities) was strengthened.
- The use and control of the externally developed elements of the food safety management program were clarified.
- The concept of documented information was introduced.
- The concept of a hazard control plan was introduced.

ISO 22000 integrates the principles of the HACCP system and application steps developed by the Codex Alimentarius Commission. By means of auditable requirements, it combines the HACCP plan with prerequisite programs. Hazard analysis is the key to an effective food safety management system, since conducting a hazard analysis assists in organizing the knowledge required to establish an effective combination of control measures. ISO 22000 requires that all hazards that may be reasonably expected to occur in the food chain, including hazards that may be associated with the type of process and facilities used, are identified and assessed. Thus, it provides the means to determine and document why certain identified hazards need to be controlled by a particular organization and why others do not.

During hazard analysis, the organization determines the strategy to be used to ensure hazard control by combining the prerequisite programs and the HACCP plan. HACCP is a management system in which food safety is addressed through the analysis and control of biological, chemical, and physical hazards from raw material production, procurement, and handling, to manufacturing, distribution, and consumption of the finished product. A HACCP system helps you focus on the hazards that affect food safety through hazard identification and establish critical control limits at critical points during the production process.

HACCP is a systematic approach to the identification, evaluation, and control of food safety hazards based on the following seven principles:

Principle 1: Conduct a hazard analysis

Principle 2: Determine the CCPs

Principle 3: Establish critical limits

Principle 4: Establish monitoring procedures

Principle 5: Establish corrective actions

Principle 6: Establish verification procedures

Principle 7: Establish record-keeping and documentation procedures

A more detailed description of HACCP is included in Chapter 6.

FDA Food Safety Modernization Act: Risk-based Food Safety Prevention

The FDA Food Safety Modernization Act (FSMA), the most important reform of U.S. food safety laws in more than 70 years, was signed into law by President Obama on January 4, 2011 It aims to ensure that the U.S. food supply is safe by shifting the focus from responding to contamination to preventing it.

Prevention as the Cornerstone of the FDA Food Safety Modernization Act. About 48 million people (one in six Americans) get sick, 128,000 are hospitalized, and 3,000 die each year from foodborne diseases, per recent data from the Centers for Disease Control and Prevention (CDC). This is a significant public health burden that is largely preventable.

The FSMA enables FDA to better protect public health by strengthening the food safety system. It enables FDA to focus more on preventing food safety problems rather than relying primarily on reacting to problems after they occur. Building a new food safety system based on prevention will take time, and FDA is creating a process for getting this work done. Congress has established specific implementation dates in the legislation. The funding the Agency gets each year, which affects staffing and vital operations, will affect how quickly FDA can put this legislation into effect. FDA is committed to implementing the requirements through an open process with opportunity for input from all stakeholders.

The following are among FDA's key new prevention authorities and mandates. For the first time, FDA will have a legislative mandate to require comprehensive, science-based preventive controls across the food supply. This mandate includes:

- *Mandatory preventive controls for food facilities*. Food facilities are required to implement a written preventive controls plan. This involves:

1. Evaluating the hazards that could affect food safety and specifying what preventive steps, or controls, will be put in place to significantly minimize or prevent the hazards

2. Specifying how the facility will monitor these controls to ensure they are working
3. Maintaining routine records of the monitoring
4. Specifying what actions the facility will take to correct problems that arise

- *Mandatory produce safety standards*. FDA must establish science-based, minimum standards for the safe production and harvesting of fruits and vegetables. Those standards must consider naturally occurring hazards, as well as those that may be introduced either unintentionally or intentionally, and must address soil amendments (materials added to the soil such as compost), hygiene, packaging, temperature controls, animals in the growing area, and water.
- *Authority to prevent intentional contamination*. FDA must issue regulations to protect against the intentional adulteration of food, including the establishment of science-based mitigation strategies to prepare and protect the food supply chain at specific vulnerable points.

The FDA Food Safety Modernization Act (FSMA) Preventive Controls for Human Food rule is now final, and compliance dates for some businesses began in September 2016. This final rule is the product of an unprecedented level of outreach by FDA to industry; consumer groups; the agency's federal, state, local, and tribal regulatory counterparts; academia; and other stakeholders. This outreach began before the rule was proposed in January 2013.

In response to input received during the comment period and during hundreds of engagements that included public meetings, webinars, listening sessions, and visits to farms and food facilities across the country, FDA issued a supplemental notice of proposed rulemaking in September 2014. The proposed revisions were designed to make the originally proposed rule more practical, flexible, and effective for industry, while still advancing FDA's food safety goals.

The final rule has elements of both the original and supplemental proposals, in addition to new requirements that are the outgrowth of public input received during the comment period for both proposals. For example, flexibility was built into key requirements, including control of the supply chain, and the definition of farms—which are exempt from these regulations—was significantly changed to reflect modern farming practices.

Key Requirements. Covered facilities must establish and implement a food safety system that includes an analysis of hazards and risk-based preventive controls. The rule sets requirements for a written food safety plan that includes:

- *Hazard analysis*. The first step is hazard identification, which must consider known or reasonably foreseeable biological, chemical, and physical hazards. These hazards could be present because they occur naturally, are unintentionally introduced, or are intentionally introduced for economic gain (if they affect the safety of the food).
- *Preventive controls*. These measures are required to ensure that hazards requiring a preventive control will be minimized or prevented. They include process, food allergen, and sanitation controls, as well as supply chain controls and a recall plan.
- *Oversight and management of preventive controls*. The final rule provides flexibility in the steps needed to ensure that preventive controls are effective and to correct problems that may arise.
- *Monitoring*. These procedures are designed to provide assurance that preventive controls are consistently performed. Monitoring is conducted as appropriate to the preventive control. For example, monitoring of a heat process to kill pathogens would include actual temperature values and be more frequent than the monitoring of preventive maintenance activities used to minimize metal hazards, which could be a simple record of the date on which the activity took place.
- *Corrective actions and corrections*. Corrections are steps taken to timely identify and correct a minor, isolated problem that occurs during food production. Corrective actions include actions to identify a problem with preventive control implementation, reduce the likelihood the problem will reoccur, evaluate affected food for safety, and prevent it from entering commerce. Corrective actions must be documented with records.
- *Verification*. These activities are required to ensure that preventive controls are consistently implemented and effective. They include validating with scientific evidence that a preventive control is capable of effectively controlling an identified hazard; calibration (or accuracy checks) of process monitoring and verification instruments such as thermometers; and reviewing records to verify that monitoring and corrective actions (if necessary) are being conducted. Product testing and environmental monitoring are

possible verification activities but are only required as appropriate to the food, facility, nature of the preventive control, and role of that control in the facility's food safety system. Environmental monitoring generally would be required if contamination of a ready-to-eat food with an environmental pathogen is a hazard requiring a preventive control.

- *Current good manufacturing practices (cGMPs).* These were updated and clarified. Among the more significant changes are the following:
  - The final rule does not include nonbinding provisions, which are more appropriate for guidance.
  - Some of the previously nonbinding provisions, such as education and training, are now binding.
  - Management is required to ensure that all employees who manufacture, process, pack, or hold food are qualified to perform their assigned duties.
  - Such employees must have the necessary combination of education, training, and/or experience necessary to manufacture, process, pack, or hold clean and safe food. Individuals must receive training in the principles of food hygiene and food safety, including the importance of employee health and hygiene.
  - FDA's long-standing position that cGMPs address allergen cross-contact is now explicit in the regulatory text.

Supply Chain: Risk Consideration for Regulated Products

Globalization presents serious challenges to the ability of the U.S. medical product safety system to protect American families. Among these challenges are the growth in the number of medical products produced abroad, the increasingly complex path that medical products travel from source materials to consumers, and the greater chance for the intentional substitution of ingredients for profit (economic adulteration).

FDA is implementing a comprehensive approach to medical product safety along the supply chain, like its approach to food safety. The goal of these efforts is to protect Americans from contamination, diversion, counterfeiting, and other risks that could harm patients. In FDA's fiscal year (FY) 2009 appropriation, Congress directed FDA "to prepare and provide to the Committees on Appropriations a comprehensive approach to ensuring the safety of medical products from the manufacturing of raw ingredients or components to consumer use." The response to this congressional inquiry was the Report on FDA's *Approach to Medical Product Supply Chain Safety* presented July 14, 2009.

For the purposes of this report, medical products include human drugs, biologics, and medical devices, but not special dietary products, medicated feeds, veterinary drugs, or radiation-emitting products that are not otherwise medical devices. Other reports on medical product safety underscore the importance of FDA's supply chain safety efforts. In December 2007, FDA's Science Board Subcommittee on Science and Technology issued a report concluding that FDA was not positioned to meet the demands of its regulatory responsibilities considering the challenges of regulating globalized industries and the need to manage new science and technologies. The 2007 report recommended that FDA strengthen its collaboration with other government agencies to improve medical product safety programs. A supplemental report, issued in May 2008 by the Science Board Subcommittee of the Office of Regulatory Affairs (ORA), contained similar findings and recommended that FDA conduct more domestic and foreign inspections coupled with certification programs for foreign manufacturers, greater inspection by foreign regulatory authorities, and improved risk-management inspection procedures. FDA has incorporated these approaches into its supply chain safety initiative.

Supply chain safety refers to minimizing risks that arise anywhere along the supply chain continuum, from sourcing a product's raw materials, ingredients, and components through the product's manufacture, importation, sale, and distribution. Although FDA has a significant number of efforts under way that address medical product safety issues beyond the supply chain, addressing supply chain safety presents some of the greatest challenges. This report identifies trends and challenges that FDA faces all along the supply chain. To address them and protect public health, FDA has adopted four primary strategies:

1. Focus on prevention
2. Enhance regulatory information
3. Improve FDA's scientific and analytic capabilities
4. Expand risk-based inspection and enforcement

The report describes these four strategies and the steps that FDA is taking to implement them. The report concluded that:

> Providing patients with safe and effective medical products is a vital part of FDA's mission to protect and promote the public health.

> FDA is committed to implementing a modern medical product safety system that focuses on prevention, provides enhanced regulatory information to FDA, is supported by improved FDA scientific and analytic capabilities, and expands risk-based inspections and enforcement. FDA has already taken many steps in furtherance of its comprehensive approach to improving medical product safety; we have implemented a wide range of new initiatives and instituted policies that are aimed at making sure that American consumers receive only safe medical products.
>
> At the same time, FDA recognizes that its approach must continue to evolve and remain flexible, as challenges to the medical product supply chain will also continue to evolve.
>
> Although we have made significant progress, much remains. We look forward to working with Congress and our other partners on additional ways to maintain a modern safety system so that Americans have the safe and effective medical products they rightly deserve.

FDA also identified as part of the report the following areas of opportunity to increase safety of the medical products supply chain:

- Increase foreign and domestic medical product inspections and other field activities conducted by ORA
- Improve laboratory infrastructure to increase the volume of sample analysis
- Provide technical support to FDA foreign offices and field operations conducting foreign and domestic device manufacturing inspections
- Provide on-site training to foreign regulators on medical device safety and quality standards
- Create and translate computerized training classes on medical device safety and quality standards into several languages
- Verify registration of medical device establishments
- Develop and issue regulations on a unique device identification system
- Increase technical support for foreign and domestic inspections and enhance risk analysis tools for biologic products

- Develop quality systems for product testing and lot release of biological products
- Strengthen science and computing to ensure that FDA can adequately regulate new drug manufacturing technologies
- Identify and improve enforcement against internet sites that expose consumers to unapproved drug products and fraudulently marketed drug products
- Modernize and enhance information technology systems to collect, store, and analyze regulatory, scientific, and risk-based information necessary to ensure the safety and effectiveness of medical products

Supply chain safety is a shared responsibility of not only the producer but also the country of origin, the importer, the importing country, and the final company in the supply chain. Companies that sell contaminated products because of loose supply chain oversight need to face serious penalties and cannot excuse themselves by blaming their suppliers.

Companies will need to have a strong supply chain management system in order to withstand the increased scrutiny by FDA. Companies will need to start to review their supply chain management program now, identify gaps, and address them before an FDA inspection. Additionally, those companies that mostly use contract manufacturers to produce and distribute their products will need to assure that they have tight control over those contractors since it appears that FDA will now hold them responsible.

Food and medical product companies will need to ensure that their supplier management programs are compliant, maintain oversight, and facilitate strong supplier relationships. They should also ensure that their suppliers are evaluated and that regulations and guidance are followed according to the product or service they provide. Companies should focus on several areas in their supplier programs, such as:

- Knowledge about their supply chains, including participants and vulnerabilities
- Risk management of the supply chain
- Qualification of prospective suppliers
- Audits and testing programs based on identified risks
- Quality and performance of suppliers
- Building relationship with suppliers, such as partnerships, alliances, and so on

More and more, FDA is observing companies that do not have control over their supply chain. These companies need to ensure compliance and begin treating their suppliers as an extension of their own company.

The drug-shortage crisis in the United States is not just a supply chain issue but also a quality-management crisis with life-and-death implications. These crises are driven by poor supply chain coordination and market failure. Shortages of pharmaceutical drugs pose a serious and growing threat to public health. Drug shortages are becoming more severe as well as more frequent. The affected medicines include cancer treatments, anesthesia drugs, and other drugs that are critical to the treatment and prevention of serious diseases and life-threatening conditions.

Endnotes

1. https://www.accessdata.fda.gov/scripts/cder/rems/index.cfm.
2. Annex VIII of the EU MDR 2017/745 details the different classification rules.

3

Principles of Quality Risk Management

KEY PRINCIPLES

ISO 31000:2018 describes eight key principles that view risk management as an elementary process of generating success and value for the organization. These eight principles can be regarded as the basic and essential qualities required for risk management. They are:

1. *Integrated*. Risk management is not a stand-alone activity that is separate from the main activities and processes of the organization. Risk management is a responsibility of an organization's management and an integral part of all organizational processes, including strategic planning and all project and change management processes. It is a part of decision-making in every department and embedded into the organization's processes.
2. *Structured and Comprehensive*. Approaching risk management in a systematic way contributes to efficiency and consistent results within the organization as well as a comprehension for everyone involved. Risk management is structured with guidelines and procedures to follow in order to maintain productivity and efficacy.
3. *Customized*. Risk management processes are not one-size-fits-all and must be tailored to the organization's external and internal context in order to reach objectives. When the context is established in both internal and external environments, objectives can be captured and risk management can be customized to the unique organization.

4. *Inclusive*. The involvement of stakeholders allows their knowledge and views to be considered, guaranteeing that risk management is relevant and up to date. Risk management is transparent; it is easy to understand and doesn't include confusing jargon, allowing stakeholders to be included in the framework.

5. *Dynamic*. Context and knowledge within an organization change constantly and should be acknowledged as they do. Risk management must respond to change continually and in a timely manner to maintain efficiency and results. Risks emerge, change, and disappear as internal and external events occur, so risk management must be anticipatory.

6. *Best Available Information*. An organization will never have all the information needed, but action must be taken when an organization has the best available data. Historical and current information, as well as the limitations of these, must be taken into account. All known information should be available to stakeholders.

7. *Human and Cultural Factors*. Risk management is influenced significantly by human behavior and culture. The organization's capabilities, as well as the goals of the people within and around it, must be recognized by risk management to achieve, or inhibit, the goals of the business.

8. *Continual Improvement*. Improving continually through experience ensures the organization's resiliency. PDCA is a risk management process: plan, do, check, act. This is a cycle that keeps the organization continually improving while factors change over time. Appropriately adapting to results in risk management allows the organization to grow exponentially in every aspect and continue to do so.

These eight principles are the foundation for managing risk and are considered when creating processes in all areas of an organization with the ISO 31000 standard. Without the foundation provided by these developed principles, the framework of ISO 31000 risk management would not be sound. The best risk management uses these principles to manage uncertainty for an organization to reach its objectives and continue achieving goals. When an organization manages risk while incorporating the eight principles, it will see consistent and reliable results.

Regarding this same topic, the Product Quality Research Institute Manufacturing Technical Committee (PQRI–MTC) commissioned a study

(Frank et al. 2011) to summarize actual risk management practices. The outcome of this study, in terms of the main quality risk management principles used by pharmaceutical and biotechnology industry, is:

1. Compliance with applicable laws: Risk assessment should be used to assess how to ensure compliance and to determine the resulting prioritization for action—not for a decision regarding the need to fulfill applicable regulations or legal requirements.
2. Risk can only be effectively managed when it is identified, assessed, considered for further mitigation, and communicated. This principle embodies the four stages of an effective quality risk management process as defined by ICH Q9: risk assessment (risk identification, analysis, and evaluation), risk control (risk reduction and acceptance), risk communication, and risk review.
3. All quality risk evaluations must be based on scientific and process-specific knowledge and ultimately linked primarily to the protection of the patient. Risk assessment is based on a strong understanding of the underlying science, applicable regulations, and related processes involved with the risk under analysis. Collectively, these components should be assessed first and foremost with regard to the potential impact to the patient.
4. Effective risk management requires a sufficient understanding of the business, the potential impact of the risk, and ownership of the results of any risk management assessment.
5. Risk assessment must consider the probability of a negative event in combination with the severity of that event. This principle also serves as a useful working definition for risk (risk represents the combination of the probability and severity of any given event).
6. It is not necessary or appropriate to always use a formal risk management process. Rather, the use of an informal risk management process (for example, an empirical assessment) is acceptable for areas that are less complex and have lower potential risk. Risk decisions are made by industry every day. The complexity of the events surrounding each decision and the potential risk involved are important inputs in determining the appropriate risk assessment methodology and corresponding level of analysis required. For less complex, less risky decisions, a qualitative analysis (decision tree) of the options may be all that is required.

To be effective and efficient, organizations should aim at the appropriate level of performance of their risk management framework in line with the criticality of the decisions that are to be made. The list of attributes below represents a high level of performance in managing risk. To assist organizations in measuring their own performance against these criteria, some indicators are given for each attribute. The key outcomes are that the organization has a *current, correct,* and *comprehensive understanding of its risks,* and the organization's risks are within its risk criteria.

Continual Improvement

An emphasis is placed on continual improvement in risk management through the setting of organizational performance goals, measurement, review, and the subsequent modification of processes, systems, resources, capability, and skills. This risk management performance assessment is an integral part of the overall organization's performance assessment and measurement system for departments and individuals.

Full Accountability for Risks

Enhanced risk management includes comprehensive, fully defined, and fully accepted accountability for risks, controls, and risk treatment tasks. Designated individuals fully accept accountability, are appropriately skilled, and have adequate resources to check controls, monitor risks, improve controls, and communicate effectively about risks and their management to external and internal stakeholders. The organization ensures that those who are accountable are equipped to fulfill that role by providing them with the authority, time, training, resources, and skills sufficient to assume their accountabilities.

Application of Risk Management in All Decision-Making

All decision-making within the organization, whatever the level of importance and significance, involves the explicit consideration of risks and the application of risk management to some appropriate degree. This can be indicated by records of meetings and decisions showing that explicit discussions on risks took place. In addition, it should be possible to see that all components of risk management are represented within key processes for decision-making in the organization, for example, for decisions on the allocation of capital, on major projects, and on restructuring and organizational changes. For these reasons, soundly based risk management is seen within the organization as providing the basis for effective governance.

Continual Communications

Enhanced risk management includes continual communications with external and internal stakeholders, including comprehensive and frequent reporting of risk management performance, as part of good governance. This can be indicated by communication with stakeholders as an integral and essential component of risk management. Communication is rightly seen as a two-way process such that properly informed decisions can be made about the level of risks and the need for risk treatment against properly established and comprehensive risk criteria.

Full Integration into the Organization's Governance Structure

Risk management is viewed as central to the organization's management processes, such that risks are considered in terms of the effect of uncertainty on objectives. The governance structure and process are based on the management of risk. Effective risk management is regarded by managers as essential for the achievement of the organization's objectives.

This section can be fairly summarized into the following two primary principles of quality risk management:

1. The evaluation of the risk to quality should be based on scientific knowledge and ultimately link to the protection of the patient.
2. The level of effort, formality, and documentation of the quality risk management process should be commensurate with the level of risk.

SCIENCE AND RISK-BASED APPROACH TO PRODUCT QUALITY

As medical products manufacturing evolves from an art to a science- and engineering-based activity, application of this enhanced science and engineering knowledge in regulatory decision-making, establishment of specifications, and evaluation of manufacturing processes should improve the efficiency and effectiveness of both manufacturing and regulatory decision-making. As a pillar of the initiative (and guiding principle), regulators must ensure that science-based policies and standards form the foundation upon which product quality regulation is based. One cannot forget that people (the American public, in the case of FDA) are the ultimate customers of medical product manufacturing companies, and because the public is often unable to judge the quality of a product, the goal of the regulatory system is to make sure patients do not have to worry about the quality of their medicines and medical products.

A great example of this new focus is the implementation of new technologies, such as process analytical technology (PAT), which is detailed in Chapter 5. To keep pace with the many advances in manufacturing quality management and to be able to more effectively allocate its limited regulatory resources, FDA is implementing a risk-based approach to regulating pharmaceutical manufacturing. The approach is applied to the review, compliance, and inspectional components of FDA regulation.

The intensity of FDA oversight needed to be related to several factors, including the degree of a manufacturer's product and process understanding and the robustness of the quality system controlling their process. For example, changes to complex products made with complex manufacturing processes may need more regulatory oversight. Process changes with critical variables that have not been sufficiently defined (as often happens with many older products) may require the submission of additional data or comparability protocols. In other cases, changes in well-understood processes could be managed under a firm's change control procedures. Additional factors in performing risk-based quality assessments include instances when manufacturing processes are crucial to the safety of the product (as a sterilization process for sterile products) or when products serve a critical medical need or have a critical public health impact (such as products for the prevention of communicable diseases).

Other considerations, such as public health impact and the compliance status or compliance history of the manufacturer, can also continue to influence the intensity of FDA oversight. Following are several FDA examples of the application of this science- and risk-based approach:

- *Prioritization of United States manufacturing site inspection.* For almost 20 years now, FDA has used a risk-based approach for prioritizing domestic manufacturing site inspections for certain human pharmaceuticals. The purpose of this approach is to help the Agency predict where its inspections are likely to achieve the greatest public health impact. The frequency and/or scope of inspections should be reduced for firms that FDA determines have acquired sufficient process understanding and have succeeded in implementing effective quality systems approaches. FDA expectation was also to create positive incentives for other firms to implement effective quality systems at their manufacturing sites. At the same time, the Agency started to apply risk-based principles to the product quality review process.

- *Risk-based approach for clinical investigations.* In August 2011, FDA released a draft guidance entitled *Oversight of Clinical Investigations: A Risk-Based Approach to Monitoring.* FDA published the draft guidance to assist sponsors of clinical investigations in developing risk-based monitoring strategies and plans for clinical investigations of human drug and biological products, medical devices, and combinations thereof. The guidance is intended to make clear that sponsors can use a variety of approaches to meet their monitoring responsibilities during clinical investigations. The guidance describes a risk-based approach to monitoring that focuses on critical study parameters and relies on a combination of monitoring activities to effectively oversee a study. For example, the guidance encourages greater use of centralized monitoring methods where appropriate. The guidance also makes recommendations about how to develop monitoring plans and document monitoring activities (Food and Drug Administration 2011b).
- *New process validation guidance.* In January 2011, FDA published the guidance *Process Validation: General Principles and Practices,* and it followed a risk-based approach. This is consistent with the direction of previous drafts. The new guidance is also consistent with ICH Q5A and Q10 and European guidelines. It confirms that process validation should be looked at as a lifecycle activity rather than a one-time event. FDA says the guidance "encourages the use of modern pharmaceutical development concepts, quality risk management, and quality systems at all stages of the manufacturing process lifecycle...."

As with these other guidelines, FDA's final guidance focuses on a risk-based approach and does not include specific recommendations. For instance, those expecting it to specify a number of *process performance qualification* batches required for licensure will undoubtedly be disappointed. Despite industry requests for clarification on this point, the guidance recommends the use of risk-based decision-making during performance qualification to assess process consistency, and acknowledges that risk management tools should be used during all stages of process validation. While many would have liked to see FDA endorse the *de facto* standard of "three consecutive batches" for process performance qualification, the agency's decision to leave this out allows companies flexibility in interpreting this requirement. If process validation work conducted during the design phase is done properly, three process performance qualification batches should be sufficient to demonstrate process consistency. Expanded characterization and sampling during the verification stage further reduces the risk of this approach.

The new guidance confirms the evolution of process validation from a one-time activity performed just prior to regulatory submissions to a life-cycle activity that begins early in development and continues long after commercialization. According to FDA, the guidance "... encourages the use of modern pharmaceutical development concepts, quality risk management, and quality systems at all stages of the manufacturing process lifecycle," an approach that has been used for decades in other industries to improve product quality, lower costs, and decrease time to market. It is now time for the biopharmaceutical industry to embrace and implement these concepts to improve their drug development process and decrease regulatory risk.

- *Quality Management Maturity (QMM) Program*. FDA's Center for Drug Evaluation and Research (CDER) is establishing a program to promote quality management maturity (QMM) at drug manufacturing establishments. The QMM program aims to encourage drug manufacturers to implement quality management practices that go beyond current cGMP requirements. The goals of this program are fourfold:

  1. Foster a strong quality culture mindset
  2. Recognize establishments that have advanced quality management practices and acknowledge establishments that strive to continually improve quality management practices
  3. Identify areas where quality management practices can be enhanced and provide suggestions for growth opportunities
  4. Minimize risks to product availability to assure reliable market supply

Adopting mature quality management practices supports a more reliable drug supply chain by reducing the occurrence of quality-related failures and improving the ability of establishments to maintain performance during expected and unexpected supply chain disruptions. Integrating business and manufacturing operations with quality practices and technological advancements can help achieve higher levels of maturity. This can optimize manufacturing process performance and product quality, enhance supply chain reliability, and foster proactive continual improvement.

QUALITY SYSTEMS APPROACH

Best practices in quality management methods, particularly in other high-tech industries, have undergone significant progress since 1978 when the drug cGMP regulations were updated. FDA wanted to ensure that its regulatory practices encourage similar progress in the pharmaceutical industry. The guidance for industry *Quality Systems Approach to Pharmaceutical Current Good Manufacturing Practice*, published in September 2006 (Food and Drug Administration 2006), describes a comprehensive quality systems model that manufacturers could use and highlights the model's consistency with the cGMP regulations for manufacturing human and veterinary drugs, including biological products. The guidance explains how manufacturers implementing such a comprehensive quality system can ensure that they comply fully with the cGMP regulations (21 CFR parts 210 and 211).

This guidance describes a comprehensive quality systems model that, if implemented, will allow manufacturers to support and sustain robust, modern quality systems that are consistent with cGMP regulations. The guidance demonstrates how and where the elements of this comprehensive model can fit within the requirements of the cGMP regulations. The inherent flexibility of the cGMP regulations should enable manufacturers to implement a quality system in a form that is appropriate for their specific operations. The overarching philosophy articulated in both the cGMP regulations *and* in robust modern quality systems is:

> Quality should be built into the product, and testing alone cannot be relied on to ensure product quality.

This guidance is intended to serve as a bridge between the 1978 regulations and our current understanding of quality systems. A quality system addresses the public and private sectors' mutual goal of providing a high-quality drug product to patients and prescribers. A well-built quality system should reduce the number of (or prevent) recalls, returned or salvaged products, and defective products entering the marketplace.

It is important that the cGMP regulations are harmonized to the extent possible with other widely used QMSs, including ISO 9000, non–U.S. pharmaceutical quality management requirements, and FDA's own medical device quality system regulations. This guidance serves as a first step in highlighting common elements between the cGMP regulations and QMS. With the globalization of pharmaceutical manufacturing and the increasing prevalence of drug–device and biologic–device combination products, the convergence of quality management principles across different regions and among various product types is very desirable.

FDA has concluded that modern quality systems, when coupled with manufacturing process and product knowledge and the use of effective risk management practices, can handle many types of changes to facilities, equipment, and processes without the need for prior approval regulatory submissions. Manufacturers with a robust quality system and appropriate process knowledge can implement many types of improvements. In addition, an effective quality system, by lowering the risk of manufacturing problems, may result in shorter and fewer FDA inspections.

A quality system can provide the necessary framework for implementing quality by design, continual improvement, and risk management in the drug manufacturing process. A quality system adopted by a manufacturer can be tailored to fit the specific environment, taking into account factors such as scope of operations, complexity of processes, and appropriate use of finite resources. The following key concepts are used throughout the guidance as they relate to the manufacture of pharmaceutical products:

- *Quality*. Every pharmaceutical product has established identity, strength, purity, and other quality characteristics designed to ensure the required levels of safety and effectiveness. For the purposes of this guidance document, the phrase *achieving quality* means achieving these characteristics for a product.

- *QbD and product development*. QbD means designing and developing a product and associated manufacturing processes that will be used during product development to ensure that the product consistently attains a predefined level of quality at the end of the manufacturing process. QbD, in conjunction with a quality system, provides a sound framework for the transfer of product knowledge and process understanding from drug development to the commercial manufacturing processes and for post-development changes and optimization. The cGMP regulations, when viewed in their entirety, incorporate the concept of quality by design.

- *Quality risk management*. Quality risk management is a valuable component of an effective quality systems framework. Quality risk management can, for example, help guide the setting of specifications and process parameters for drug manufacturing, assess and mitigate the risk of changing a process or specification, and determine the extent of discrepancy investigations and corrective actions.

- *CAPA (corrective and preventive action).* CAPA is a well-known cGMP regulatory concept that focuses on investigating, understanding, and correcting discrepancies while attempting to prevent their recurrence. Quality system models discuss CAPA as three separate concepts, all of which are used in this guidance:

    1. Remedial corrections of an identified problem
    2. Root cause analysis with corrective action to help understand the cause of the deviation and potentially prevent recurrence of a similar problem
    3. Preventive action to avert recurrence of a similar potential problem

- *Change control.* Change control is another well-known cGMP concept that focuses on managing change to prevent unintended consequences. The cGMP regulations provide for change control primarily through the assigned responsibilities of the quality control unit. Certain major manufacturing changes (for example, changes that alter specifications, a critical product attribute, or bioavailability) require regulatory filings and prior regulatory approval. Effective change control activities (for example, quality planning and control of revisions to specifications, process parameters, procedures) are key components of any quality system. In this guidance, change is discussed in terms of creating a regulatory environment that encourages change toward continual improvement. This means a manufacturer is empowered to make changes subject to the regulations based on the variability of materials used in manufacturing and process improvements resulting from knowledge gained during a product's life cycle.

4

General Quality Risk Management Process

INTRODUCTION TO THE QUALITY RISK MANAGEMENT PROCESS

Quality risk management is a systematic process for the assessment, control, communication, and review of risks to the quality of a drug product across the product life cycle. The emphasis on each component of the model might vary from case to case, but a robust process will incorporate consideration of all the elements at a level of detail that is commensurate with the specific risk.

The success of risk management will depend on the effectiveness of the management framework in providing the foundations and arrangements that will embed it throughout the organization at all levels. The framework assists in managing risks effectively through the application of the risk management process at different levels and within specific contexts. The framework ensures that information about risk derived from the risk management process is adequately reported and used as a basis for decision-making and accountability at all relevant organizational levels.

The different components of the framework for managing risk and the way in which they interrelate in an iterative manner are shown in Figure 4.1. The main purpose of this framework is to assist the organization in integrating risk management into its overall management system (see ISO 31000:2018 for complete details).

Figure 4.2 depicts a typical model for the risk management process based on ICH Q9(R1).

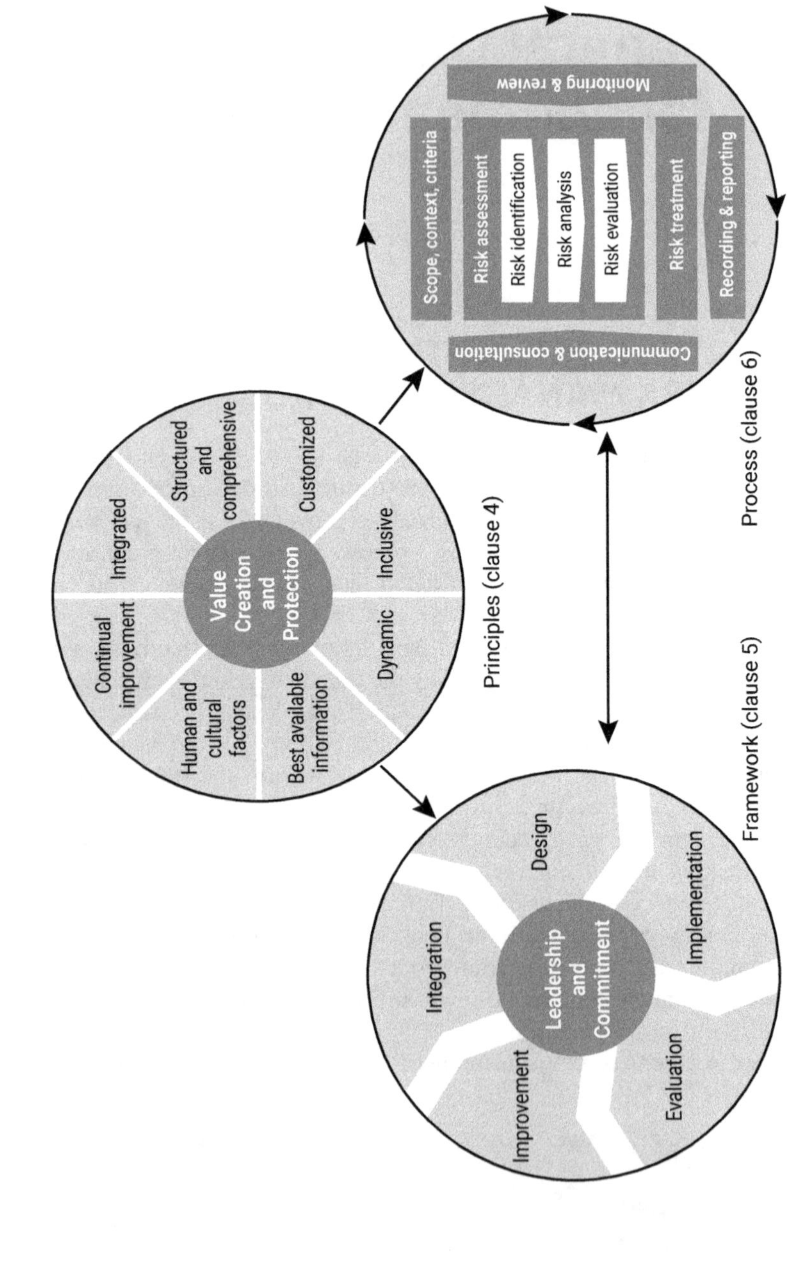

Figure 4.1 Relationships between the components of the model for managing risk.

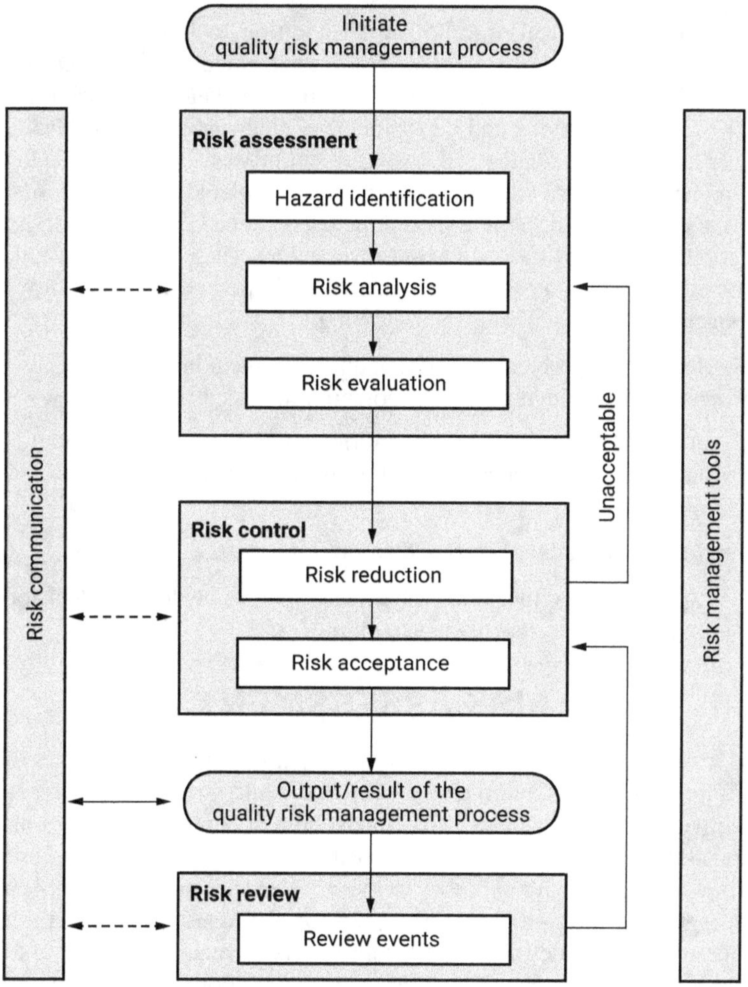

Figure 4.2 Quality risk management model.

RESPONSIBILITIES AND INITIATION OF THE RISK MANAGEMENT PROCESS

Quality risk management activities are usually undertaken by interdisciplinary teams. When teams are formed, they should include experts from the appropriate areas (for example, quality unit, business development, engineering, regulatory affairs, production operations, sales and marketing, legal, statistics, clinical, and so on) in addition to individuals who are knowledgeable about the quality risk management process. Decision-makers should take responsibility for coordinating quality risk management across various functions and departments of their organization. They also should ensure that a quality risk management process is defined, deployed, and reviewed and that adequate resources are available.

Quality risk management should include systematic processes designed to coordinate, facilitate, and improve science-based decision-making with respect to risk. Possible steps used to initiate and plan a quality risk management process include the following:

- Define the problem and/or risk question, including pertinent assumptions identifying the potential for risk
- Assemble background information and/or data on the potential hazard, harm, or human health impact relevant to the risk assessment
- Identify a leader and critical resources
- Specify a timeline, deliverables, and appropriate level of decision-making for the risk management process

RISK ASSESSMENT

Risk assessment provides decision-makers and responsible parties with an improved understanding of risks that could affect achievement of objectives, and the adequacy and effectiveness of controls already in place. It consists of the *identification* of hazards and the *analysis* and *evaluation* of risks associated with exposure to those hazards. Risk assessment begins with a well-defined problem description or risk question. When the risk in question is well defined, an appropriate risk management tool, and the types of information that will address the risk question, will be more readily identifiable. This provides a basis for decisions about the most appropriate approach to be used to treat the risks. The output of risk assessment is an input to the decision-making processes of the organization. As an aid to clearly defining the risk(s) for risk assessment purposes, three fundamental questions are often helpful:

1. What might go wrong?
2. What is the likelihood (probability) it will go wrong?
3. What are the consequences (severity)?

Risk assessment is the overall process of risk identification, risk analysis, and risk evaluation (see Figure 4.2).

Hazard Identification

Hazard identification is the systematic use of information to identify hazards related to the risk question or problem description. Information can include historical data, theoretical analysis, informed opinions, and stakeholders' concerns. Hazard identification addresses the "What might go wrong?" question, including identifying the possible consequences. This provides the basis for further steps in the quality risk management process.

The organization should identify sources of hazard, their causes, and their potential consequences. The aim of this step is to generate a comprehensive list of hazards based on those events that might create, enhance, prevent, degrade, accelerate, or delay the achievement of objectives. Comprehensive identification is critical because a hazard that is not identified at this stage will not be included in further analysis. Once a hazard is identified, the organization should identify any existing controls such as design features, people, processes, and systems.

Identification should include hazards regardless of whether their source is under the control of the organization and even when the hazard source or cause may not be evident. Hazard identification should include the examination of the indirect effects of consequences, including cascade and cumulative effects. It should also consider a wide range of consequences even if the hazard source or cause may not be evident. As well as identifying what might happen, it is necessary to consider possible causes and scenarios that show what consequences can occur. All significant causes and consequences should be considered.

The organization should apply hazard identification tools and techniques that are suited to its objectives and capabilities, and to the risks faced. Relevant and up-to-date information is important in identifying hazards. Personnel with appropriate knowledge and training should be involved in identifying hazards.

Hazard identification methods can include:

- Evidence-based methods, examples of which are checklists and reviews of historical data
- Systematic team approaches where a team of experts follow a systematic process to identify hazards by means of a structured set of prompts or questions

- Inductive reasoning techniques such as hazard and operability (HAZOP), which is described in Chapter 6

Various supporting techniques can be used to improve accuracy and completeness in hazard identification, including brainstorming and Delphi methodology. Irrespective of the actual techniques employed, it is important that due recognition is given to human and organizational factors when identifying hazards. Hence, deviations of human and organizational factors from the expected should be included in the hazard identification process as well as "hardware" or "software" events.

Preliminary Analysis. Hazards may be screened in order to identify the most significant ones or to exclude less significant or minor ones from further analysis. The purpose is to ensure that resources will be focused on the most significant hazards.

Risk Analysis

Risk analysis is the estimation of the risk associated with the identified hazards. It is the qualitative or quantitative process of linking the likelihood of occurrence and severity of harms. In some risk management tools, the ability to detect the harm (known as *detectability*) is another factor in the estimation of risk.

Risk analysis involves developing an understanding of the risk. It provides an input into risk evaluation and into decisions on whether risks need to be treated, and on the most appropriate risk treatment strategies and methods. Risk analysis can also provide an input into making decisions where choices must be made and the options involve different types and levels of risk.

Risk analysis involves consideration of the causes and sources of risk, their consequences, and the likelihood that those consequences can occur. Factors that affect consequences and likelihood should be identified. Risk is analyzed by determining consequences and their likelihood, and other attributes of the risk. An event can have multiple consequences and affect multiple objectives. Existing controls or barriers and their effectiveness and efficiency should also be included in this analysis.

The way in which consequences and likelihood are expressed and the way in which they are combined to determine a level of risk should reflect the type of risk, the information available, and the purpose for which the risk assessment output is to be used. These should all be consistent with the risk criteria. It is also important to consider the interdependence of different risks and their sources. The confidence in the determination of the level of risk and its sensitivity to preconditions and assumptions should be considered in the analysis and communicated effectively to decision-makers and, as appropriate, other stakeholders. Factors such as divergence of opinion among experts, uncertainty, availability, quality, quantity, and ongoing relevance of information, or limitations on modeling should be stated and can be highlighted.

Risk analysis can be undertaken with varying degrees of detail depending on the risk, the purpose of the analysis, and the information, data, and resources available. Analysis can be qualitative, semiquantitative, or quantitative, or a combination of these, depending on the circumstances. Consequences and their likelihood can be determined by modeling the outcomes of an event or set of events, or by extrapolation from experimental studies or from available data. Consequences can be expressed in terms of tangible and intangible impacts.

More than one technique may be required for complex applications. Risk analysis normally includes an estimation of the range of potential consequences that might arise from an event, situation, or circumstance, and their associated probabilities, in order to measure the level of risk. However, in some instances, such as when the consequences are likely to be insignificant or the probability is expected to be extremely low, a single parameter estimate may be sufficient for a decision to be made. In some circumstances, a consequence can occur as a result of a range of different events or conditions, or when the specific event is not identified. In this case, the focus of risk assessment is on analyzing the importance and vulnerability of the system components with a view to defining treatments that relate to levels of protection or recovery strategies.

Methods used in analyzing risks can be qualitative, semiquantitative, or quantitative. The degree of detail required will depend on the particular application, the availability of reliable data, and the decision-making needs of the organization. Some methods and the degree of detail of the analysis may be prescribed by legislation. Qualitative assessment defines consequence, probability, and level of risk by significance levels such as "high," "medium," and "low," may combine consequence and probability, and evaluates the resultant level of risk against qualitative criteria. Semiquantitative methods use numerical rating scales for consequence and probability, and combine them to produce a level of risk using a formula. Scales may be linear or logarithmic or have some other relationship; formulae used can also vary.

Quantitative analysis estimates practical values for consequences and their probabilities, and produces values of the level of risk in specific units defined when developing the context. Full quantitative analysis may not always be possible or desirable due to insufficient information about the system or activity being analyzed, lack of data, influence of human factors, and so on, or because the effort of quantitative analysis is not warranted or required. In such circumstances, a comparative semiquantitative or qualitative ranking of risks by specialists, knowledgeable in their respective fields, may still be effective. In cases where the analysis is qualitative, there should be a clear explanation of all the terms employed, and the basis for all criteria should be recorded.

Even where full quantification has been carried out, it needs to be recognized that the levels of risk calculated are estimates. Care should be taken to ensure that they are not attributed a level of accuracy and precision inconsistent with the accuracy of the data and methods employed. Levels of risk should be expressed in the most suitable terms for the type of risk and in a form that aids risk evaluation. In some instances, the magnitude of a risk can be expressed as a probability distribution over a range of consequences.

Controls Assessment. The level of risk will depend on the adequacy and effectiveness of existing controls.

Questions to be addressed include:

- What are the existing controls for a particular risk?
- Are those controls capable of adequately treating the risk so it is controlled to a level that is tolerable?
- Are the controls operating in the manner intended and can they be demonstrated to be effective when required?

The level of effectiveness for a particular control, or suite of related controls, may be expressed qualitatively, semiquantitatively, or quantitatively. In most cases, a high level of accuracy is not warranted. However, it may be valuable to express and record a measure of risk control effectiveness so judgments can be made on whether effort is best expended in improving a control or providing a different risk treatment.

Consequence analysis determines the nature and type of impact that could occur assuming that a particular event, situation, or circumstance has occurred. An event may have a range of impacts of different magnitudes and affect a range of different objectives and different stakeholders. The types of consequences to be analyzed and the stakeholders affected will have been decided when the context was established. Consequence analysis can vary from a simple description of outcomes to detailed quantitative modeling or vulnerability analysis.

Impacts may have a low consequence but high probability, or a high consequence and low probability, or some intermediate outcome. In some cases, it is appropriate to focus on risks with potentially very large outcomes, as these are often of greatest concern to managers. In other cases, it may be important to analyze both high- and low-consequence risks separately. For example, a frequent (chronic) but low-impact problem may have large cumulative or long-term effects. In addition, the treatment actions for dealing with these two distinct kinds of risks are often quite different, so it is useful to analyze them separately.

Consequence analysis can involve:

- Taking into consideration existing controls to treat the consequences, together with all relevant contributory factors that have an effect on the consequences
- Relating the consequences of the risk to the original objectives
- Considering both immediate consequences and those that may arise after a certain time has elapsed, if this is consistent with the scope of the assessment
- Considering secondary consequences, such as those impacting associated systems, activities, equipment, or organizations

Risk Estimation: Severity and Likelihood Analysis. Several methods can be used to estimate risk. Quantitative risk estimation is preferable when suitable data are available. In other cases, without suitable data, qualitative methods of risk estimation can be used. Risk estimation is based on the combination of the following two components:

- The probability of occurrence of harm
- The consequences of that harm (how severe it might be)

A third element frequently considered is the *detectability* of the harm situation. In the medical product manufacturing arena, this third element is carried out through numerous tests and analyses performed on the product prior to its release to customers. Severity levels and likelihood estimation must be part of a formal risk assessment procedure, which will help guide the risk assessment process and provide consistency throughout the process.

Severity Analysis. To categorize the severity of the potential harm, appropriate descriptors must be developed for potential risk and related consequences. Tables 4.1 and 4.2 provide examples of severity categorization.

Table 4.1 Basic qualitative severity level.

Category	Description
Significant/critical	Death or loss of function (inoperative medical device or drug without active ingredient)
Moderate	Reversible or minor injury
Negligible	Will not cause injury

Table 4.2 Severity categorization.

Severity	Patient safety	Regulatory compliance
Catastrophic (5)	Death	Consent decree, product seizure, cessation of operations
Critical (4)	Permanent impairment or life-threatening injury	Major observations or Warning Letter, repeated observations, product recall
Major (3)	Injury or impairment requiring professional medical intervention	Multiple minor observations
Minor (2)	Temporary injury or impairment not requiring professional medical intervention	Only minor observations or comments
Negligible (1)	Very minor inconvenience	No observations

Likelihood (Probability) Estimation. Three general approaches can be employed to estimate probability; they may be used individually or jointly:

a. *Relevant historical data* can be used to identify events or situations that have occurred in the past and hence be used to extrapolate the probability of their occurrence in the future. The data used should be relevant to the type of system, facility, organization, or activity being considered, and to the operational standards of the organization involved. If historically there is a very low frequency of occurrence, then any estimate of probability will be very uncertain. This applies especially for zero occurrences, when one cannot assume the event, situation, or circumstance will not occur in the future.

b. *Probability estimation* using predictive techniques such as fault tree analysis (see Chapter 6). When historical data are unavailable or inadequate, it is necessary to derive probability by analysis of the system, activity, equipment, or organization and its associated failure or success states. Numerical data for equipment, humans, organizations, and systems from operational experience, or published data sources, are then combined to produce an estimate of the probability of the top event. When using predictive techniques, it is important to ensure that due allowance has been made in the analysis for the possibility of common mode failures involving the coincidental failure of several different parts or components within the system arising from the same cause. Simulation techniques may be required to generate probability of equipment and structural failures due to aging and other degradation processes by calculating the effects of uncertainties.

c. *Expert opinion* can be used in a systematic and structured process to estimate probability. Expert judgments should draw on all relevant available information, including historical, system-specific, organization-specific, experimental, design, and so on. There are several formal methods for eliciting expert judgment that provide an aid to the formulation of appropriate questions. The methods available include the Delphi approach (described in Chapter 6), paired comparisons, category rating, and absolute probability judgments. Tables 4.3 and 4.4 show two examples of probability estimation using qualitative and quantitative indexes.

Uncertainties and Sensitivities. There are often considerable uncertainties associated with the analysis of risk. An understanding of uncertainties is necessary to interpret and communicate risk analysis results effectively. The analysis of uncertainties associated with data, methods, and models used to identify and analyze risk plays an important part in their application. Uncertainty analysis involves the determination of the variation or imprecision in the results that is the result of the collective variation in the parameters and assumptions used to define the results.

Table 4.3 Simplified example of quality probability level.

Category	Description
High	Likely to happen (often, frequent)
Medium	Can happen, but not frequently
Low	Unlikely to happen (rare, remote)

Table 4.4 Quantitative probability levels.[1]

Category	Probability range
Frequent	$\geq 10^{-3}$
Probable	$< 10^{-3}$ and $\geq 10^{-4}$
Occasional	$< 10^{-4}$ and $\geq 10^{-5}$
Remote	$< 10^{-5}$ and $\geq 10^{-6}$
Improbable	$< 10^{-6}$

Uncertainty is due to a combination of incomplete knowledge about a process and its expected or unexpected variability. Typical sources of uncertainty include gaps in knowledge, gaps in pharmaceutical science and process understanding, sources of harm (for example, failure modes of a process, sources of variability), and probability of the detection of problems.

Risk Evaluation

Risk evaluation compares the identified and analyzed risk against given risk criteria, and based on this comparison, the need for control can be considered. In doing an effective risk assessment, the robustness of the data set is important because it determines the quality of the output. Revealing assumptions and reasonable sources of uncertainty will enhance confidence in this output and/or help identify its limitations.

The purpose of risk evaluation is to assist in making decisions, based on the outcomes of risk analysis, about which risks need treatment and the priority for treatment implementation.

The output of a risk assessment is either a quantitative estimate of risk or a qualitative description of a range of risk. When risk is expressed quantitatively, a numerical probability is used. Alternatively, risk can be expressed using qualitative descriptors, such as "high," "medium," or "low," which should be defined in as much detail as possible. Sometimes a risk score is used to further define descriptors in risk ranking.

Decisions should take into account the wider context of the risk and include consideration of the tolerance of the risks borne by parties other than the organization that benefits from the risk. Decisions should be made in accordance with legal, regulatory, and other requirements. In some circumstances, the risk evaluation can lead to a decision to undertake further analysis. The risk evaluation can also lead to a decision not to treat the risk in any way other than maintaining existing controls.

Decisions may include:

- Whether a risk needs treatment (be controlled)
- Priorities for treatment
- Whether an activity should be undertaken
- Which of a number of paths should be followed

The nature of the decisions that need to be made and the criteria that will be used to make those decisions were decided when establishing the context, but they need to be revisited in more detail at this stage now that more is known about the particular risks identified.

The simplest framework for defining risk criteria is a single level that divides risks that need treatment from those that do not. This gives attractively simple results but does not reflect the uncertainties involved both in estimating risks and in defining the boundary between those that need treatment and those that do not.

The decision about whether and how to treat the risk may depend on the costs and benefits of taking the risk and the costs and benefits of implementing improved controls.

A common approach is to divide risks into three bands or areas (see Figure 4.3):

a. An upper band where the level of risk is regarded as intolerable regardless of the benefits the activity may bring, and risk treatment is essential whatever its cost
b. A middle band (or "gray area") where costs and benefits are considered and opportunities balanced against potential consequences
c. A lower band where the level of risk is regarded as negligible, or so small that no risk treatment measures are needed

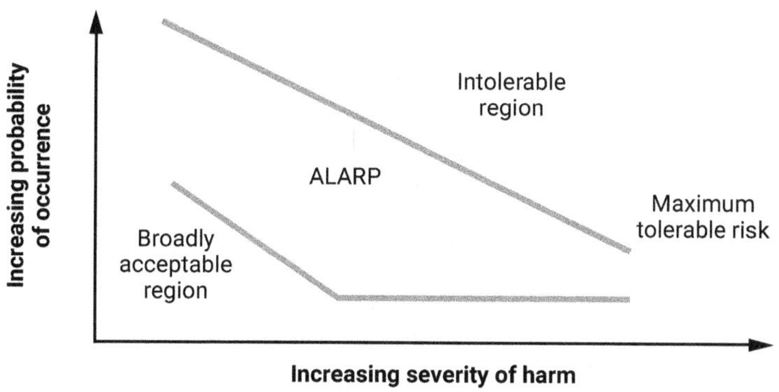

Figure 4.3 Risk estimation.

The *as low as reasonably practicable,* or ALARP, criteria system used in safety applications follows this approach: in the middle band, there is a sliding scale for low risks where costs and benefits can be directly compared, whereas for high risks, the potential for harm must be reduced until the cost of further reduction is entirely disproportionate to the safety benefit gained.

Risk Reduction "As Far as Possible" (AFAP) Compared to "As Low as Reasonably Practicable" (ALARP)

For manufacturers of medical devices, when looking to manage and reduce risk, many have looked to the ISO 14971 standard as a means of managing the risks associated with the manufacture and processing of their medical device. This standard references and suggests the use of ALARP. As a result, ALARP has become widely applied and is the basis for the acceptance by many manufacturers of the low to moderate levels of residual risks that may be associated with the development, processing, and use of their medical devices. However, under European Medical Device Regulations (MDR 2017/745) and the In Vitro Diagnostic Medical Devices Regulation (IVDR 2017/746), the requirement for those involved in development and manufacture of medical devices is that all risks need to be reduced "as far as possible."

The main difference between them is that ALARP allows a manufacturer to take economic impacts into consideration when deciding what are acceptable and unacceptable risks. AFAP does not allow economic impacts to form part of the risk decision-making process.

An important part of the risk analysis process is to ensure that you do not introduce new hazards into your quest to eliminate or minimize hazards. FDA describes its expectations about risk-based decisions in the preamble of 21 CFR Part 820. FDA states that if any risk is judged to be unacceptable, it should be reduced to acceptable levels by the appropriate means, which may include a redesign or warnings. On the other hand, the EU MDR regulation has a very strong emphasis on risk management. Risk is mentioned 243 times in the regulation compared to a combined total of only 69 times in the two original medical device directives.

ISO 14971:2019 refers you to your own risk acceptability policy to determine risk control options. Your risk policy establishes criteria for the level of control and may employ one of the following three approaches. Specifically, Clause 4.2 of ISO 14971:2019, Note 1 states that "The manufacturer's policy for establishing criteria for risk acceptability can define the approaches to risk control: reducing risk as low as reasonably practicable, reducing risk as low as reasonably achievable, or reducing risk as far as possible without adversely affecting the benefit-risk ratio." This brings manufacturers to a point where they can manage risk based on all the requirements.

European requirements (MDR and IVDR) with respect to reducing risk include:

- Reduce risk as far as possible
- Eliminate risk (or reduce as far as possible)
- Minimize risk
- Remove risk (or reduce as far as possible)
- Remove risk (or minimize as far as possible)
- Reduce risk to the lowest possible
- Avoid risk to the lowest possible

The ALARP concept has a legal interpretation that implies financial considerations. However, the European MDR does not allow financial considerations to override the General Safety and Performance Requirements for safety and performance of medical devices. If risk controls are not implemented, the justification for this must be on a basis other than financial. Instead of claiming that the residual risks are ALARP, manufacturers will need to demonstrate that risks are reduced AFAP by showing objective evidence that all possible risk control options were considered and implemented.

RISK CONTROL

Risk control includes decision-making to reduce and/or accept risks. The purpose of risk control is to reduce the risk to an acceptable level. The amount of effort used for risk control should be proportional to the significance of the risk. Decision-makers might use different processes, including benefit–cost analysis, to determine the optimal level of risk control. Risk control might focus on the following questions:

- Is the risk above an acceptable level?
- What can be done to reduce or eliminate risks?
- What is the appropriate balance between benefits, risk, and resources?
- Are new risks created as a result of the identified risks being controlled?

Table 4.5 includes some examples of risk control activities.

Table 4.5 Examples of risk control activities.

Product/process	Hazard	Inherent safe features and protective measures	Information for safety
Single-use medical device (sterile syringe)	Contamination (biological and cross-contamination)	Sterilization pouch, single use	Warning on labeling against reuse
Over-the-counter product	Criminal contamination	Tamper-evident packaging	Warning on labeling regarding tamper-evident features

Risk Reduction

Risk reduction focuses on processes for mitigating or avoiding quality risk when it exceeds a specified (acceptable) level. Risk reduction might include actions taken to mitigate the severity and probability of harm. Processes that improve the detectability of hazards and quality risks might also be used as part of a risk control strategy. The implementation of risk reduction measures can introduce new risks into the system or increase the significance of other existing risks. Hence, it might be appropriate to revisit the risk assessment to identify and evaluate any possible change in risk after implementing a risk reduction process.

Risk treatment involves selecting one or more options for modifying risks, and implementing those options. Once implemented, treatments provide or modify the controls. Risk treatment involves a cyclical process of:

- Assessing a risk treatment
- Deciding whether residual risk levels are tolerable
- If not tolerable, generating a new risk treatment
- Assessing the effectiveness of that treatment

Selection of Risk Treatment Options. Selecting the most appropriate risk treatment option involves balancing the costs and efforts of implementation against the benefits derived, with regard to legal or regulatory requirements. Several treatment options can be considered and applied either individually or in combination. The organization can normally benefit from the adoption of a combination of treatment options. When selecting risk treatment options, the organization should consider the values and perceptions of stakeholders and the most appropriate ways to communicate with them. Where risk treatment options can have an impact on risk in other units of the organization or on stakeholders, these should be involved in the decision.

Though equally effective, some risk treatments can be more acceptable to some stakeholders than to others. The treatment plan should clearly identify the priority order in which individual risk treatments should be implemented. Risk treatment itself can introduce risks. A significant risk can be the failure or ineffectiveness of the risk treatment measures. Monitoring needs to be an integral part of the risk treatment plan to give assurance that the measures remain effective. Risk treatment can also introduce secondary risks that need to be assessed, treated, monitored, and reviewed. Treatment plans should be integrated with the management processes of the organization and discussed with appropriate stakeholders. Decision-makers and other stakeholders should be aware of the nature and extent of the residual risk after risk treatment. The residual risk should be documented and subjected to monitoring, review and, where appropriate, further treatment. Table 4.6 depicts several examples of hazards and corresponding risk reduction control measures when using medical devices.

Table 4.6 Examples of hazards when using medical devices and risk reduction/mitigation measures.

Hazards	Example of recommended mitigation measures
Maintenance-related failure of equipment	Schedule maintenance
Bleeding or bruising at percutaneous insertion site	Labeling information
Inadequate instructions or procedures	Redesign instructions/procedures and improve training
Equipment use error	Develop handling/use guidelines, educate, and monitor
Local tissue injury due to material breakage	Performance bench testing (tensile strength, durability) and labeling information
Unpredictable equipment failure	Obtain more-reliable equipment and/or obtain backup
Infection	Sterilization of the devices and labeling information
Adverse tissue reaction to device, including foreign body reaction, inflammation, and so on	Biocompatibility testing

Risk Acceptance

Risk acceptance is a decision to accept risk. Risk acceptance can be a formal decision to accept the residual risk or it can be a passive decision in which residual risks are not specified. For some types of harms, even the best quality risk management practices might not eliminate risk. In these circumstances, it might be agreed that an appropriate quality risk management strategy has been applied and that the quality risk is reduced to a specified (acceptable) level. This (specified) acceptable level will depend on many parameters and should be decided on a case-by-case basis.

RISK DOCUMENTATION AND COMMUNICATION

Risk communication is the sharing of information about risk and risk management between the decision-makers and others. Parties can communicate at any stage of the risk management process (see Figure 4.2, dashed arrows). The output/result of the quality risk management process should be appropriately communicated and documented (see Figure 4.2, solid arrows). Communications might include those among interested parties (for example, regulators and industry, industry and the patient, within a company, industry, or regulatory authority). The included information might relate to the existence, nature, form, probability, severity, acceptability, control, treatment, detectability, or other aspects of risks to quality. Communication need not be carried out for each risk acceptance. Between the industry and regulatory authorities, communication concerning quality risk management decisions might be effected through existing channels, as specified in regulations and guidances.

As with any important activity related to medical products, risk management activities must be traceable.

Decisions concerning the creation of records should consider:

- The organization's needs for continuous learning
- Benefits of reusing information for management purposes
- Costs and efforts involved in creating and maintaining records
- Legal, regulatory (for example retention period), and operational needs for records

The risk assessment process should be documented together with the results of the assessment. Risks should be expressed in understandable terms, and the units in which the level of risk is expressed should be clear. The extent of the report will depend on the objectives and scope of the assessment. Content of the report must be described within a standard procedure, and, except for very simple assessments, the documentation should include:

- Objectives and scope
- Description of relevant parts of the system and their functions
- The situation, system, or circumstances being assessed
- Risk criteria applied and their justification
- Assessment methodology
- Risk identification results
- Data, assumptions, and their sources and validation
- Risk analysis results and their evaluation
- Critical assumptions (including uncertainty analysis) and other factors that need to be monitored
- Discussion of results, and conclusions and recommendations

If the risk assessment supports a continuing risk management process, it should be performed and documented in such a way that it can be maintained throughout the life cycle of the system, organization, equipment, or activity. The assessment should be updated as significant new information becomes available and the context changes, in accordance with the needs of the management process.

RISK MONITORING AND EFFECTIVENESS REVIEW

Risk management should be an ongoing part of the quality management process. A mechanism for reviewing or monitoring events should be implemented. The output/results of the risk management process should be reviewed to take into account new knowledge and experience. Once a quality risk management process has been initiated, that process should continue to be used for events that might impact the original quality risk management decision whether these events are planned (for example, results of product review, inspections, audits, change control) or unplanned (for example, root cause from failure investigations, recall). The frequency of any review should be based on the level of risk. Risk review might include reconsideration of risk acceptance decisions previously taken.

As part of the risk management process, risks and controls should be monitored and reviewed on a regular basis to verify the following:

- Assumptions about risks remain valid
- Assumptions on which the risk assessment is based, including the external and internal context, remain valid
- Expected results are being achieved
- Results of risk assessment are in line with actual experience
- Risk assessment techniques are being properly applied
- Risk treatments are effective

Accountability for monitoring and performing reviews should be established as part of a formal procedure. The risk assessment process will highlight context and other factors that might be expected to vary over time and that could change or invalidate the risk assessment. These factors should be specifically identified for ongoing monitoring and review so the risk assessment can be updated when necessary. Data to be monitored to refine the risk assessment should also be identified and collected. The effectiveness of controls should also be monitored and documented to provide data for use in risk analysis. Accountabilities for creating and reviewing the evidence and documentation should be defined. Progress in implementing risk treatment plans provides a performance measure. The results can be incorporated into the organization's overall performance management, measurement, and external and internal reporting activities.

APPLICATION OF RISK ASSESSMENT DURING LIFE-CYCLE PHASES

Many activities, projects, and products can be considered to have a life cycle, starting from initial concept and definition through realization to final completion, which might include decommissioning and disposal of hardware. Risk assessment can be applied at all stages of the life cycle and is usually applied many times with different levels of detail to assist in the decisions that need to be made at each phase. Life-cycle phases have different requirements and need different techniques. For example, during the design and development phase, when an opportunity is identified, risk assessment may be used to decide whether to proceed or not. Where several options are available, risk assessment can be used to evaluate alternative concepts to help decide which provides the best balance of positive and negative risks.

During the design and development phase, risk assessment contributes to:

- Ensuring that system risks are tolerable
- The cost-effectiveness studies
- Identifying risks impacting subsequent life-cycle phases

As the activity proceeds, risk assessment can be used to provide information to assist in developing procedures for normal and emergency conditions.

Risk management activities should begin as early as possible in the design and development phase, when it is easier to prevent problems, rather than correcting them later. For each identified hazard, the risk in both normal and fault conditions is estimated. In risk evaluation, the manufacturer decides whether risk reduction is needed. The results from this risk evaluation, such as the need for risk control measures, then become part of the design input. Risk control measures are part of the design output and are evaluated during design verification. This design input/output/verification cycle will iterate and continue throughout the overall design control process until the residual risks have been reduced to an acceptable level and can be maintained at an acceptable level. The overall effectiveness of risk control measures is confirmed during design validation. Relying exclusively on design and development processes to control risk is not sufficient. Even the best design and development processes can fall short of ensuring error-free design output.

Once the medical product reaches the market, risk management activities should be linked to quality management processes, for example, production and process controls, CAPA, customer feedback, and servicing in the case of some medical devices.

ENHANCED RISK MANAGEMENT

All organizations should aim at the appropriate level of performance of their risk management framework in line with the criticality of the decisions that are to be made. The list of attributes below represents a high level of performance in managing risk. To assist organizations in measuring their own performance against these criteria, some tangible indicators are given for each attribute. Following is a list of attributes of enhanced risk management.

Continual Improvement

An emphasis is placed on continual improvement in risk management through the setting of organizational performance goals, measurement, review, and the subsequent modification of processes, systems, resources, capability, and skills. This can be indicated by the existence of explicit performance goals against which the organization's and individual manager's performance is measured. There will be a periodic review of performance and then a revision of processes, and the setting of revised performance objectives for the following period.

Full Accountability for Risks

Enhanced risk management includes comprehensive, fully defined, and fully accepted accountability for risks, controls, and risk treatment tasks. Designated individuals fully accept accountability, are appropriately skilled, and have adequate resources to check controls, monitor risks, improve controls, and communicate effectively about risks and their management to external and internal stakeholders. The organization ensures that those who are accountable are equipped to fulfill that role by providing them with the authority, time, training, resources, and skills sufficient to assume their accountabilities.

Application of Risk Management in All Decision-Making

All decision-making within the organization, whatever the level of importance and significance, involves the explicit consideration of risks and the application of risk management to some appropriate degree. Soundly based risk management is seen within the organization as providing the basis for effective governance.

Continual Communications

Enhanced risk management includes continual communications with external and internal stakeholders, including comprehensive and frequent reporting of risk management performance, as part of good governance.

Full Integration in the Organization's Governance Structure

Risk management is viewed as central to the organization's management processes, such that risks are considered in terms of the effect of uncertainty on objectives. The governance structure and process are based on the management of risk. Effective risk management is regarded by managers as essential for the achievement of the organization's objectives.

Endnote

1. Taken from ISO/TR 24971:2020, Table 5.

5

Integration of Risk Management with Quality Management

Quality risk management is a process that supports science-based and practical decisions when integrated into quality systems. As outlined in the introduction, appropriate use of quality risk management does not obviate industry's obligation to comply with regulatory requirements. However, effective quality risk management can facilitate better and more informed decisions, can provide regulators with greater assurance of a company's ability to deal with potential risks, and might affect the extent and level of direct regulatory oversight. In addition, quality risk management can facilitate better use of resources by regulators and manufacturers.

Training in quality risk management processes provides for greater understanding of decision-making processes and builds confidence in quality risk management outcomes. Quality risk management must be integrated into existing operations and documented appropriately. Chapter 7 provides examples of situations in which the use of the quality risk management process might provide information that could then be used in a variety of medical product operations. A common understanding and application of quality risk management principles will facilitate mutual confidence and promote more-consistent decisions among different countries' regulators on the basis of the same information. This collaboration is a critical step in the development of global policies and guidelines that integrate and support quality risk management practices.

RISK MANAGEMENT PLANS

ISO 31073:2022 defines a *risk management plan* (RMP) as a "scheme within the risk management framework specifying the approach, the management components and resources to be applied to the management of risk." Furthermore, it says that "management components typically include procedures, practices, assignment of responsibilities, sequence and timing of activities" and also that "the risk management plan can be applied to a particular product, process and project, and part or whole of the organization." Risk management activities must be planned. Therefore, for the particular medical product being considered, the manufacturer must establish and document a risk management plan in accordance with the risk management process. The risk management plan must be part of the risk management file.

This plan must include at least the following:

a. The scope of the planned risk management activities, identifying and describing the medical product and the life-cycle phases for which each element of the plan is applicable

b. Assignment of responsibilities and authorities

c. Requirements for review of risk management activities

d. Criteria for risk acceptability, based on the manufacturer's policy for determining acceptable risk, including criteria for accepting risks when the probability of occurrence of harm cannot be estimated

e. Method to evaluate the overall residual risk and criteria for acceptability of the overall residual risk based on the manufacturer's policy for determining acceptable risk

f. Verification activities

g. Activities related to the collection and review of relevant production and post-production information.

If the plan changes during the life cycle of the medical product, a record of the changes must be maintained in the risk management file. Compliance is checked by inspection of the risk management file. The RMP can be a separate document or it can be integrated within other documentation, for example, within quality management system documentation. The makeup and level of detail for the plan should be commensurate with the level of risk associated with the medical product. The requirements identified below should be considered as the minimum requirements for an RMP. Manufacturers can include other items such as time schedule, risk analysis tools, or a rationale for the choice of specific risk acceptability criteria.

Scope of the Plan

The scope identifies and describes the medical product and the life-cycle phases for which each element of the plan is applicable. All elements of the risk management process should be mapped to the manufacturer's defined product life cycle. Some of the elements of the risk management process will occur during the phases of the manufacturer's established product realization process, such as design and development control. The remaining elements will occur during the other life-cycle phases through to product decommissioning. The risk management plan provides this mapping for a specific product either explicitly or by reference to other documents. Although all the risk management activities need to be planned, a manufacturer can have several plans covering different parts of the life cycle.

Assignment of Responsibilities and Authorities

The plan should identify the personnel with responsibility for the execution of specific risk management activities, for example, reviewers (including independent ones not directly involved in the process under evaluation), subject matter experts, and individuals with approval authority.

Requirements for Review of Risk Management Activities

The plan should detail how and when these risk management reviews will occur for the specific medical product. The requirements for the review of risk management activities could be part of other quality system review requirements (for example, design and development review activities).

Criteria for Risk Acceptability

Criteria for risk acceptability (benefit-risk analysis) are derived from the manufacturer's policy for determining acceptable risk. The criteria can be common for similar categories of medical product. Criteria for risk acceptability can be part of the manufacturer's established QMS, which can be referenced in the plan.

An important element when establishing criteria for risk acceptability is a benefit–risk determination. In the United States over the past several years, FDA has been taking steps to develop an enhanced, structured approach to benefit–risk assessment in regulatory decision-making for medical products, starting with human drugs, biologic products, and medical devices. This structured approach was named the *benefit–risk assessment framework,* and it is discussed in Chapter 7.

Verification Activities

The plan will specify how two distinct verification activities will be carried out. The first verification is required to demonstrate that the risk control measure has been implemented. The second verification is required to ensure that the implemented measure is effective (it actually reduces the risk). Verifying the effectiveness of risk control measures can require the collection of different types of data (manufacturing data, clinical data, usability studies in the case of medical devices, and so on). The plan can detail the verification activities explicitly or by reference to the plan for other verification activities.

Methods of Obtaining Relevant Post-Production Information

The method of obtaining post-production information can be part of established QMS procedures (for example, monitoring and measuring). Manufacturers must establish procedures to collect information from various sources such as medical staff, patients, service personnel, training personnel, incident reports, and customer feedback. While a reference to the QMS procedures can suffice in most cases, any product-specific requirements should be directly added to the risk management plan. The plan should include documentation of decisions, based on a risk analysis, about what sort of post-market surveillance is appropriate for the medical product, for example, whether reactive surveillance is adequate or whether proactive studies are needed. Pre- and post-market risks are discussed later in this chapter.

Example of Risk Management Plan Applied to Medication Guides

In many countries, special medicine safety plans are required as part of a medicine's approval process and for maintenance of approval. These requirements are specified in such documents as:

- ICH Harmonized Tripartite Guideline *Pharmacovigilance Planning—E2E* (2004)
- The European Medicines Agency's (EMA) *Guidance on the format of the risk management plan (RMP) in the EU* (2018)

- FDA Guidance for Industry—*Development and Use of Risk Minimization Action Plans* (RiskMAPs 2005), which in 2007 evolved into a new statutorily-based safety plan called a Risk Evaluation and Mitigation Strategy (REMS).[1]

Under the FDA Amendments Act of 2007, FDA may require a REMS to be developed to ensure that the benefits of a medicine outweigh its risks. This could be required as part of the approval process for a new product or for an approved product when new safety information emerges. Some medicines that were previously approved with RiskMAPs are now deemed to have a REMS. A REMS is a strategy to manage a known or potential serious risk associated with a medicine.

Its purpose is to allow patients continued access to certain medicines for which there are safety concerns that may be managed through appropriate use. All REMSs must include a timetable for assessment of the REMS.

A REMS may also include a medication guide or patient package insert and a communication to healthcare practitioners explaining elements to assure safe use of the product, such as:

- Special training or certification for healthcare professionals who prescribe or dispense the medicine
- Dispensing of the medicine only in certain settings (for example, in a hospital) or with evidence of safe use conditions
- Monitoring or registration of each patient using the medicine

Prior to the enactment of such legislation and regulations, pharmaceutical companies (recognizing that clinical studies do not necessarily represent "real-world" life experience) often designed RMPs in the preapproval stages of developing new medicines and have included them in submissions for approval. In Europe, risk management plans are now routinely required by EMA as part of the medicine approval process. An RMP includes a summary of important identified risks of the medicine, potential risks, and missing information. This summary incorporates the safety profile of the medicine at that time in its life cycle, either during preclinical testing, preapproval clinical development, or post-approval. The RMP specifies data on known and potential safety risks (drawn from preclinical and/or clinical study results for the medicine), as well as areas of risk that have not yet been studied or not studied extensively, such as specific (or larger) patient populations, patients with other medical conditions or in other treatment settings, and interactions with other medication. Pharmacovigilance and risk management activities that might be included in an RMP fall into two categories:

a. Routine activities that would generally be conducted for any medicine at the same stage of development where no special safety concerns have arisen
b. Additional activities designed to address identified safety concerns

Routine pharmacovigilance would include the safety evaluations incorporated in clinical trials and the monitoring and reporting of spontaneous adverse events post-approval. Routine risk management activities would include ensuring that suitable warnings are included with all product information and careful labeling/packaging of the medicine. In an RMP the action plan might include calling for additional pharmacovigilance in the form of:

- Active surveillance (for example, medical records reviews, patient or physician interviews, prescription event monitoring, data from disease or drug exposure registries)
- Epidemiology studies (retrospective or prospective)
- Further clinical studies (specific safety studies, larger studies over longer periods)
- Drug utilization studies that describe how a drug is marketed, prescribed, and used in a specified population (often stratified by age, gender, concomitant medications, or other factors), and how these factors influence clinical, social, and economic outcomes

If the action plan specifies additional risk minimization activities, these could include:

- Additional educational material about the medicine and its use, such as patient information brochures, visual aids, physician prescribing guides/checklists, pharmacist dispensing guides/ checklists, healthcare provider letters, and so on.
- Training programs (patient- or physician-oriented).
- Restricted use of the medicine. For example, for use or dispensing only in a hospital or where specific equipment (for example, resuscitation equipment) is available, availability only in limited unit sizes, and so on.

RISK-BASED CAPA SYSTEM

The CAPA system[2] is a risk-based control mechanism. A well-designed and effectively implemented CAPA system provides a reliable source of quality data in a timely manner.

Risk management concepts have been part of the medical device world for many years. At first, the regulators used the term "hazard analysis," and it was part of the HACCP methodology. For the last decade or so, HACCP terminology has been restricted to food safety, and FDA, ISO, and other regulators embraced the term *risk analysis,* which evolved to the current broader term of *risk management.* Risk analysis requirements are incorporated only into the design control (§820.30) element of FDA's QSR, but the preamble of this regulation includes mentions about risk analysis expectations across many elements. The international standard (ISO 14971:2019) establishes risk management requirements to determine the safety of a medical device by the manufacturer during the product life cycle.

For other regulated products, such as drugs, the application of risk management concepts was more recently done in the form of an international guidance document, ICH Q9(R1) *Quality Risk Management,* which was adopted as a nonbinding guidance by the drug and biologic centers of FDA in June 2006 and revised in 2023.

Having in mind those risk management principles, typical questions must be answered: Do I need to document every deviation/nonconformity? Do I always need corrective and preventive actions? How soon must companies fix their CAPA problems? CAPA and risk management are two interlocked concepts that cannot be separated. All our decisions regarding CAPA must be filtered throughout the risk management system. Let's now answer these questions.

Do I need to document every deviation or nonconformity? The answer is yes. Every time you detect some kind of "problem" it is necessary to look into it and document it. Repeating the primary concept of the CAPA system, continuous improvement requires analysis of the issue to discover its root cause before you can implement actions to prevent its reoccurrence. To be able to fix the cause of the problem, you must first discover its causes. Without some kind of investigation or evaluation, the probability is low that you can reach the real root causes. However, resources are not unlimited (actually, they are becoming more and more scarce) and, definitively, not all issues have the same significance. As the QSR preamble states, "at times a very in-depth investigation will be necessary, while at other times a simple investigation, followed by trend analysis or other appropriate tools, will be acceptable." Therefore, you must prioritize, and risk assessment is one of the best tools you can use for this purpose. The significance of the product or quality issue can be evaluated by considering, for example, the criteria described in Table 5.1.

Table 5.1 CAPA risk assessment criteria.

Criterion	Categories and examples
Does it have the potential for a patient or user safety issue?	• *Critical or catastrophic:* can cause death or significant disability to patient or user (contaminated injectable drug, critical drug mix-up, contaminated catheter) • *Marginal:* can cause minor injuries to patient or user (overpotent or subpotent drugs, incorrect diagnoses) • *Negligible:* no injury to patient or user (cosmetic defect, empty box without product)
Type or classification of the product	• Device class I • Device class II • Device class III • Intravenous drug or sterile product • Drug other than intravenous (oral, cutaneous) • Drug with narrow therapeutic ranges
Does it affect the reliability, effectiveness, or usability of the product? Note: Consider the worst case	• Totally affected: not working, not usable, or not effective (missing product, broken device) • Partially affected: underfill, low count/quantity • Not affected
Does the issue cause the product to fall outside of established specifications?	• Final specification failure • Nonfinal specification failure • Acceptance specification • Validity (system suitability) specification
Does it affect the labeling of the product?	• Final label incorrect (lot number, expiration date) • Nonfinal label incorrect

Continued

Table 5.1 *Continued.*

Criterion	Categories and examples
How frequent is the problem?	• First time occurrence • Occasional • Frequent
Has the frequency of occurrence of the issue changed?	• Improving • Worsening
How difficult is it to detect the issue?	• Not detectable (customer detected it) • Detected by chance (shipping operator detected it) • Detected by process (inspection detected the failure)
Does it represent a regulatory risk (that is, can product be considered as adulterated or misbranded)?	• Mix-ups • Product distributed prior to completing its record review

Most regulated companies perform some kind of risk evaluation based on the frequency and the severity (importance, significance) of the event. Situations in which the frequency is rare and the severity is low may not require further investigation. Nevertheless, this evaluation must be documented. In other words, if you can demonstrate (with objective evidence) that the problem has low frequency and no significant danger, then you could pass on this investigation and focus your effort on more significant issues.

The biggest concern with this evaluation is that a vast majority of regulated companies focused the severity evaluation exclusively on the safety of the patient. Based on that evaluation, they assigned very low risk scores to deviations and nonconformances that represent major violations of cGMPs and therefore render such products as adulterated.

Do I always need corrective and preventive actions? If you investigated and discovered the root causes of a problem, it would be insane not to fix them. FDA's position on this matter can be found in the preamble to the October 7, 1996, Medical Devices QSR. In comment 159 of the preamble, which relates to the degree of corrective or preventive actions, FDA states, "FDA cannot dictate in a regulation the degree of action that should be taken because each circumstance will be different, but FDA does expect the manufacturer to develop procedures for assessing the risk, the actions that need to be taken for different levels of risk, and how to correct or prevent the problem from recurring, depending on that risk assessment."

A remarkable observation on this side of the CAPA system is that many companies always require both corrective and preventive actions, even in situations where no true preventive action can be applied. In some cases, the reason to require them is simply that the CAPA form includes both types of actions, and therefore, both are always required.

How soon must companies fix their CAPA problems? The third question refers to the timeliness of failure investigations and corrective or preventive actions. Time frames for completing the different CAPA actions must be established based on the risk of the situation under investigation. This constitutes one of the biggest opportunities for improving CAPA systems. One simple approach used by several companies is to complete investigations (that is, the root cause investigation) in four weeks for low-risk situations, three weeks for medium-risk situations, and two weeks for high-risk situations.

In the cases previously mentioned, risk classification is normally based on frequency and severity alone. Our recommendation is to use risk management criteria to determine how deeply and how quickly every nonconformance or deviation should be treated. These risk criteria must be clearly defined in written procedures. One example might be establishing who is responsible for evaluating product or quality issues and determining whether a failure investigation is necessary. Another example would be maintaining a record when no failure investigation is made, including the reason and the name of the individual responsible for the decision. The procedure should also determine the depth to which a failure investigation is to be carried out and when an investigation should not pursue corrective action.

Table 5.2 depicts a simple way to carry out this task by segregating nonconformities and deviations into three categories based on the previously described risk criteria. For each situation, an overall risk score is determined by considering the worst-case scenario of the eight dimensions under analysis. Check marks indicate the risk classification that is assigned to each dimension. For example, if the problem can have a critical or catastrophic impact on the safety of the patient, then its risk score must be high, independent of any other dimension such as product classification, problem detectability, and so on. Table 5.2 can be applied to processes and systems, including equipment failure where no product was directly affected.

Table 5.3 is an example of how to use this risk assessment score matrix for an issue affecting a class III medical device.

Table 5.2 Risk assessment score matrix.

Criterion	Categories	Risk score		
		Negligible or low (1)	**Medium (2)**	**High (3)**
Safety	Critical or catastrophic			✓
	Marginal		✓	
	Negligible	✓		
Product classification	Device Class I	✓		
	Device Class II	✓		
	Device Class III		✓	
	Intravenous drug or sterile product		✓	
	Drug with narrow therapeutic ranges		✓	
	Other drug products	✓		
Reliability or effectiveness	Totally affected			✓
	Partially affected		✓	
	Not affected	✓		
Product specification	Final specification failure			✓
	Nonfinal specification failure		✓	
	Specifications are not affected	✓		
Product labeling	Final product labels			✓
	Nonfinal product labels		✓	
	No labeling is affected	✓		
Frequency or trending	First-time occurrence (isolated event)	✓		
	Occasional but improving	✓		
	Occasional but worsening		✓	
	Frequent		✓	

Continued

Table 5.2 *Continued.*

Criterion	Categories	Risk score		
		Negligible or low (1)	**Medium (2)**	**High (3)**
Detectability	Not detectable or not detected			✓
	Detected by chance		✓	
	Detected by the regular process	✓		
Regulatory risk	Product can be considered adulterated or misbranded			✓
	Product is not adulterated nor misbranded	✓		

Table 5.3 Example of risk assessment.

Criterion	Categories	Risk score		
		Negligible or low (1)	**Medium (2)**	**High (3)**
Safety	Critical or catastrophic			✓
	Marginal		✓	
	Negligible	(✓)		
Product classification	Device Class I	✓		
	Device Class II	✓		
	Device Class III		(✓)	
	Intravenous drug or sterile product		✓	
	Drug with narrow therapeutic ranges		✓	
	Other drug products	✓		
Reliability or effectiveness	Totally affected			✓
	Partially affected		✓	
	Not affected	(✓)		

Continued

Table 5.3 *Continued.*

Criterion	Categories	Risk score		
		Negligible or low (1)	**Medium (2)**	**High (3)**
Product specification	Final specification failure			✓
	Nonfinal specification failure		✓	
	Specifications are not affected	(✓)		
Product labeling	Final product labels			✓
	Nonfinal product labels		✓	
	No labeling is affected	(✓)		
Frequency or trending	First-time occurrence (isolated event)	(✓)		
	Occasional but improving	✓		
	Occasional but worsening		✓	
	Frequent		✓	
Detectability	Not detectable or not detected			✓
	Detected by chance		✓	
	Detected by the regular process	(✓)		
Regulatory risk	Product can be considered adulterated or misbranded			✓
	Product is not adulterated nor misbranded	(✓)		

A medium risk score was determined for the example in Table 5.3. Now you can use this risk score to determine the content and priority level of each nonconformance investigation (NCI), as detailed in Figure 5.1 and Figure 5.2, which describe the characteristics of each nonconformance investigation type.

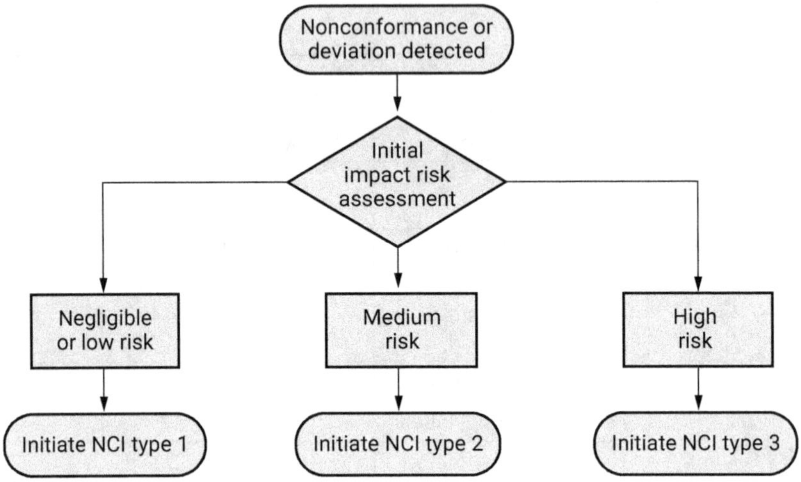

Figure 5.1 Risk prioritization of investigations.

Nonconformance Investigation (NCI)		
Type 1	**Type 2**	**Type 3**
• Only negligible or low-risk scores are obtained • One week to complete • Document the event and the corrections taken • Monthly track and trending of type 1 NCI	• At least one dimension had a medium-risk score • *30 days* to complete • Document the event, root cause analysis, and corrections taken • Need to generate a CAPA plan	• At least one dimension had a high-risk score • *20 days* to complete • Document the event, root cause analysis, and corrections taken • Need to generate a CAPA plan

Note: Complete NCI means document approval.

Figure 5.2 Types of nonconformance investigations.

EXAMPLES OF INTEGRATION BETWEEN RISK MANAGEMENT AND OTHER QUALITY MANAGEMENT TOOLS

This section covers the following three examples of such integration:

- The process analytical technology (PAT) approach
- Risk-based inspection site selection
- The quality system inspection technique (QSIT)

The Process Analytical Technology (PAT) Approach

FDA's PAT (Food and Drug Administration 2004b) initiative is an effort to facilitate the introduction of new manufacturing technologies in the pharmaceutical industry for achieving more efficient processes. *Process analytical technologies* are systems that enhance process understanding and assist in identifying and controlling critical points in a process. These include appropriate measurement devices that can be placed at-, in-, or on-line; statistical and information technology tools; and a scientific systems approach for data analysis to control processes to ensure production of in-process materials and final products of desired quality.

PAT has been defined by FDA as a mechanism for designing, analyzing, and controlling pharmaceutical manufacturing processes through the measurement of critical process parameters that affect critical quality attributes of the medical product. The PAT initiative is one of the elements within the broader FDA initiative of Pharmaceutical cGMPs for the 21st Century: A Risk-Based Approach. FDA has actively involved stakeholders in the development of this initiative. The goal of PAT is to understand and control the manufacturing process, which is consistent with the basic quality principle that *quality cannot be tested into products; it should be built-in by design*. Quality is built into pharmaceutical products through a comprehensive understanding of:

- The intended therapeutic objectives, patient population, route of administration, and pharmacological, toxicological, and pharmacokinetic characteristics of a drug
- The chemical, physical, and biopharmaceutic characteristics of a drug
- The design of a product and selection of product components and packaging based on drug attributes
- The design of manufacturing processes using principles of engineering, material science, and quality assurance to ensure acceptable and reproducible product quality and performance throughout a product's shelf life

Using this approach of building quality into products, this guidance highlights the necessity for process understanding and opportunities for improving manufacturing efficiencies through innovation and enhanced scientific communication between manufacturers and the Agency. Increased emphasis on building quality into products allows more focus to be placed on relevant multifactorial relationships between material, manufacturing process, environmental variables, and their effects on quality.

This enhanced focus provides a basis for identifying and understanding relationships between various critical formulation and process factors and for developing effective risk mitigation strategies (for example, product specifications, process controls, and training). The data and information useful in helping understand these relationships can be leveraged through preformulation programs and development and scale-up studies, as well as from improved analysis of manufacturing data collected over the life of a product.

Effective innovation in development, manufacturing, and quality assurance would be expected to better answer questions such as the following:

- What are the mechanisms of degradation, drug release, and absorption?
- What are the effects of product components on quality?
- What sources of variability are critical?
- How does the process manage variability?

The FDA initiative can facilitate innovation in development, manufacturing, and quality assurance by focusing on process understanding. These concepts are applicable to all manufacturing situations. A desired goal of the PAT framework is to design and develop processes that can consistently ensure a predefined quality at the end of the manufacturing process. Such procedures would be consistent with the basic tenet of quality by design and could reduce risks to quality and regulatory concerns while improving efficiency. Gains in quality, safety, and/or efficiency will vary depending on the product and are likely to come from:

- Reducing production cycle times by using on-, in-, and/or at-line measurements and controls
- Preventing rejects, scrap, and reprocessing
- Considering the possibility of real-time release
- Increasing automation to improve operator safety and reduce human error
- Facilitating continuous processing to improve efficiency and manage variability
  - Using small-scale equipment (to eliminate certain scale-up issues) and dedicated manufacturing facilities
  - Improving energy and material use and increasing capacity

Risk-Based Inspection Site Selection

The Federal Food, Drug, and Cosmetic Act states that FDA is to inspect domestic drug manufacturing establishments at least once every two years (Food and Drug Administration 2004b). However, the Agency was unable to attain this statutory requirement mainly due to a lack of resources and a lack of an adequate prioritization tool. As part of the initiative Pharmaceutical cGMPs for the 21st Century: A Risk-Based Approach, FDA began a risk-based inspection pilot for prioritizing drug manufacturing establishments for routine inspection. The model was based on a risk-ranking and filtering method with the ultimate objective of making the best use of its limited surveillance and enforcement resources while maximizing the impact of those resources on public health.

Prioritizing sites for inspection has been a long-standing challenge for FDA managers. Typically, FDA district offices have identified specific sites in their geographical areas for inspection each year. These decisions were made based on a variety of informally applied factors, including, for example, a district manager's knowledge of the inspectional history and corporate culture of the district, as well as the perceived risk to the public health of manufacturing errors.

Even before this initiative, FDA tried to use some kind of prioritization to better utilize inspectional resources. Three categories of facilities were identified as high priority for inspections: those that produce sterile drug products, those that produce prescription drugs, and new registrants that had not been inspected previously. Recognizing that FDA could not inspect every manufacturing site at equal frequency and depth with the present inspectional resources, a working group was created to develop a more rigorous risk-based approach to resource allocation for cGMP inspectional oversight. A risk-ranking and filtering model was developed for this task.

To develop the risk-ranking model for site selection, FDA followed a multistep analytical process like those described previously: hazard identification and conceptual modeling, risk estimation, and risk filtering.

Hazard Identification and Conceptual Modeling. A two-step process was conducted: first, a wide range of factors that could be incorporated in the model (potential risk factors) were identified. Second, these factors were organized into principal, or top-level, components of the risk-ranking model. A diverse group of FDA experts with experience in drug review, manufacturing controls, and inspections participated in a number of brainstorming sessions and a survey to identify potential risk factors. The general question posed of these experts was "In your experience, what are the principal factors important in predicting adverse impacts to drug quality?" Specific questions that the expert group considered, directly or indirectly, in identifying candidate risk factors included:

- What hazards (sources of harm) related to manufacturing can adversely impact drug quality attributes?
- What variables are associated with, or predictive of, those hazards?
- What processes and process parameters are critical for quality attributes?
- What factors may affect the identified hazards and critical parameters and processes?
- What factors are predictive of high- or low-quality manufacturing?

This process resulted in a list of more than 70 potential risk factors, of which some are potentially predictive (for example, batch sizes, dedicated lines, and so on) and others are directly relevant to quality and public health (for example, sterility or intrinsic toxicity). To develop the conceptual framework for the risk-ranking model, these factors needed to be hierarchically organized according to their level of generality and causal relationships. By direct observation, several FDA experts organized the list of risk factors into broad categories, or top-level components, of product, process, and facility (see Table 5.4).

Table 5.4 Top-level components for the site selection model.

Factor category	Description	Example(s)
Product	Factors pertaining to the intrinsic properties of drug products such that quality deficiencies could potentially and adversely impact public health	Dosage form Intrinsic chemical properties
Facility	Factors relating to characteristics of a manufacturing site believed to be predictive of potential quality risks, such as the lack of effective quality systems	Poor cGMP compliance history
Process	Factors pertaining to aspects of drug manufacturing operations that may predict potential difficulties with process control and/or vulnerability to various forms of contamination	Measuring Mixing Compression Filling

The construct for a risk-based site selection model was developed based on the above modeling framework. Under this construct, a *site risk potential* (SRP) is a function of the weighted risk potentials for each of the three categories: product, facility, and process. The risk potential for each top-level component is, in turn, a function of select potential risk factors. Thus, an SRP score is derived by a mathematical combination of weights or ranks assigned to select potential risk factors. The assignment of weights or ranks to select potential risk factors is based on either empirical evidence, expert judgment, or a combination of both. To calculate the SRP score, weights and rankings for individual potential risk factors are numerically discrete values (for example, 0.1, 1, 2, 3, 6).

Implementation of the Model and Risk Estimation. The selection of potential risk factors within each top-level component (product, facility, and process) and their weight assignments have significant influence on the SRP score. When implementing the risk-ranking model, it is important that the assignment of a weight or rank to each potential risk factor is based on empirical evidence and on systematic and transparent collection, and analyses of expert judgments. Many potential risk factors that were initially identified by the FDA expert group were excluded from the current implementation of the model because of lack of data or other data limitations, including difficulties in linking data elements with specific manufacturing sites and differing data dictionaries among various existing databases. The current iteration of the Agency's pilot risk-ranking model includes the elements depicted in Figure 5.3.

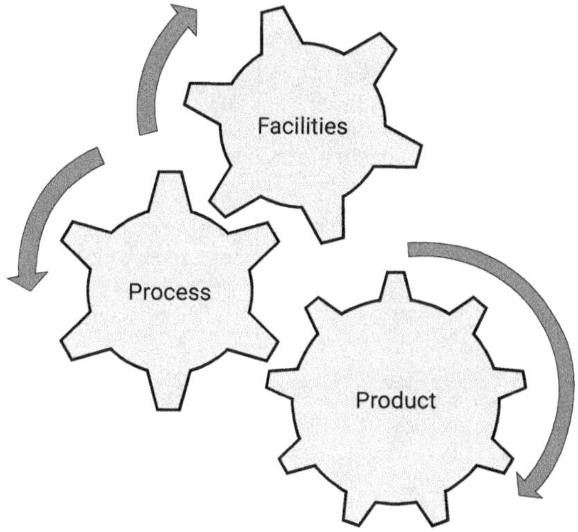

Figure 5.3 Site risk—potential elements.

Product Component. Table 5.5 depicts the two types of factors in the product component of the model.

Process Component. It is clear that some processes are more complex and more susceptible to problems than other processes. It was further recognized that one primary goal of the cGMP inspections is to ensure that processing operations are in a state of control. Thus, consensus among FDA experts was that it would be important to include process-related risk factors in the risk-ranking model. Based on the working group deliberation, two types of process-related factors were identified for inclusion in the survey and subsequently the risk-ranking model:

- Factors associated with maintaining process control
- Factors associated with potential vulnerability to product or environmental contamination

Table 5.5 Product component factors.

1. Intrinsic factors	2. Past recalls for quality defects
Factors such as sterility, medical gas, and the determination of prescription (Rx) versus over-the-counter (OTC) were used to distinguish between products with higher and lower potential for public health consequence should there be a drug defect. For example, the current model assumes that if there is a quality defect, sterile drugs would have a higher public health consequence than non-sterile drugs; hence, sterile drugs are given a higher weight. In addition, the current model considers sites manufacturing specific products where there is a heightened risk of cross-contamination, such as sites manufacturing highly sensitizing agents (for example, antibiotics) and at least one other product using similar processing methods.	FDA human drug recall data provide information on past recalls for quality defects with potential for human health hazard. The data were analyzed based on product types (for example, routes of administration and dosage forms). Using a likelihood and severity matrix, weights were assigned to product types based on frequency of recall and health hazard severity (class I, II, and III). Product types with a high frequency of recall and high hazard severity (type of recall) were given higher weights.

The survey was designed to elicit from appropriate FDA experts a risk-ranking (for example, from high to low) of the probability of loss of a state of control and, independently, the vulnerability of the process to contamination for a product category and processing operations associated with that product category.

Experts were surveyed on risks associated with commonly employed manufacturing operations (for example, measuring, mixing, compression, and filling) and for a variety of product categories (for example, immediate- and modified-release solid oral drugs, sterile liquids, metered-dose inhalers, and active ingredients produced by chemical and fermentation processes). Survey results were statistically analyzed, and process control and contamination weights were assigned to each product category identified in the expert elicitation survey.

Facility Component. The facility component of the risk-ranking model includes four factors:

1. History of violation (for example, cGMP deficiencies have higher weights)
2. History of inspection (for example, no prior inspection, newly registered/licensed, or no cGMP inspection in the past two years have higher weights than those with recent cGMP inspection)
3. Estimated volume of production output (surrogate for exposure, for example, higher volume and production output, higher weights)
4. Type of establishment (for example, manufacturer, repacker, contract lab)

These factors were identified from existing FDA databases and assigned weights using an ordinal scale similar to the *product* and *process* factors. Most of the factors are self-explanatory.

Risk Filtering and Model Summary. To test the robustness of the risk-ranking model, FDA conducted a pilot analysis of an inventory of more than 1,500 facilities. The pilot ranking, based on the historical data as described above, showed that the model could adequately spread the SRP score for filtering and that the distribution of the SRP scores was only slightly skewed. In the current iteration, the risk-ranking model will lead to a ranked list of sites for routine cGMP inspection coverage such that a site's chance of being selected for inspection will increase if it:

- Has not been inspected recently (or ever)
- Has a higher production volume

- Has a history of significant violations
- Makes products associated with a high frequency of recalls, particularly for serious defects
- Uses processes expected to have a greater potential for cross-contamination and/or loss of a state of control
- Makes sterile and/or prescription drugs

This model does not exclude any site or product type from being potentially selected as a high-priority site. It is expected that sites with effective quality systems and superior process understanding will be less likely to have a history of significant violations or have multiple product recalls, and such sites would, generally, be expected to have less frequent or intense inspectional coverage than would otherwise be the case. Similarly, sites that have successfully implemented risk mitigation techniques would likely have a better inspectional history and fewer recalls. Individual risk factor weights and final scoring information, as well as specific information used to generate the weights and scores, were intentionally not provided by FDA in order to be able to effectively enforce the law.

The Quality System Inspection Technique (QSIT)

Once the QSR was issued in 1996, FDA created a team to reengineer the inspection process used by the agency to perform quality system/good manufacturing practice inspections at medical device manufacturing facilities. The new inspection technique was called the *quality system inspection technique* (QSIT) (Food and Drug Administration 1999a). The QSIT approach to inspections was derived from the theory that there are seven subsystems in the QSR (21 CFR, Part 820). Four primary areas were chosen on which to focus the inspection: management controls, design controls, corrective and preventive actions (CAPA), and production and process controls. The remaining three subsystems are covered via "linkages" within the QSIT guide. This document provides guidance for inspecting medical device manufacturers against the QSR (21 CFR, Part 820) and related regulations.

Satellite programs are included in the QSIT due to their correlation in the inspection process with the related subsystem. The CAPA subsystem is the logical "jumping-off" point to begin inspecting for medical device reporting, corrections and removals, and medical device tracking programs that relate to a firm's post-market activities. Rather than check every aspect of the firm's quality system, the so called "top-down" subsystem approach focuses on those elements that are most important in meeting the requirements of the QSR and that are key quality indicators. Between six and 15 inspectional objectives are provided for the review of each subsystem. The review includes both a (broad) review of whether the firm has procedures in place and appears to meet the requirements and a closer (detailed) review of some records to verify that the requirements have been implemented in actual production, design, and daily quality assurance situations. This document represents an extraordinary benchmark, advising companies where to align their internal audit programs.

The QSIT guide provides guidance to FDA field staff on a new inspectional process that may be used to assess a medical device manufacturer's compliance with the QSR and related regulations. Field investigators may conduct an efficient and effective comprehensive inspection using this guidance material, which will help them focus on key elements of a firm's quality system. *The Guide to Inspections of Quality Systems* provides instructions for conducting medical device quality system/GMP inspections. It is to be used in conjunction with the compliance program entitled Inspection of Medical Device Manufacturers (7382.845).

The subsystem approach is designed to provide the inspector with the key objectives that can help determine a firm's state of compliance. The process was designed to account for the time constraints placed on field investigators when performing device quality system inspections. If FDA can focus inspection effort on key elements of a firm's quality system, both the inspector and the manufacturer can benefit from a more efficient and effective evaluation of that quality system.

When an inspector begins an inspection by looking at one or more instances of quality problems, such as nonconforming device reports, and works his/her way back through the firm's quality system, you are under a "bottom-up" inspection. This method has been helpful in zeroing in on specific problems and evaluating the firm's actions relating to those problems. However, with the "top-down" approach, FDA is looking at the firm's "systems" for addressing quality before the inspector actually looks at specific quality problems. In the top-down approach, the inspector "touches bottom" in each of the subsystems by sampling records rather than working his/her way from records review backward toward procedures.

The top-down approach begins each subsystem review with an evaluation of whether the firm has addressed the basic requirements in that subsystem by defining and documenting appropriate procedures. This is followed by an analysis of whether the firm has implemented the requirements of that subsystem. The quality system was divided into seven subsystems, along with related satellite programs. Based on discussions between the device industry and the Agency, four major subsystems that are the basic foundation of a firm's quality system were chosen. Those four major subsystems are management control, corrective and preventive action (CAPA) (with satellite medical device reporting, corrections and removals, and medical device tracking), design controls, and production and process controls (P&PC) (with satellite sterilization process controls).

The duration of inspection is related to the depth of the inspection. The guide was designed to accomplish a complete review of all four subsystems in approximately one week. While the length of each inspection will vary, using key inspectional objectives will help assure that every inspector looks at the most important elements of the firm's quality system during the inspection.

Most device firms are inspected more than once. By probing different subsystems, devices, or processes each time, FDA will eventually have covered most of the firm's quality system. You are not expected to cover everything in the firm and in the narrative each time. You are expected to evaluate the firm's quality system but also do it in an efficient and focused manner. Thus, you should limit the depth of coverage when necessary to meet the time frame suggested. As a general rule of thumb, one day should be sufficient to cover each subsystem when using the top-down approach described within this document. In practice, you may find that the inspection of a certain subsystem may take half a day, while another may take one-and-a-half days. This situation would still reflect an overall one day per subsystem time frame.

PRE- AND POST-MARKET RISKS

Regulatory expectations are that manufacturers of medical products apply good risk assessment practices during the whole life cycle of their products. Risk opportunities are everywhere, and using a life-cycle approach enables manufacturers to better understand the risks and hazards associated with their products and, more importantly, how to mitigate those risks. Life cycle includes all phases in the life of the product, from the initial development, through marketing, until the product's discontinuation.

U.S. FDA does not have a uniform approach to pre- and post-market risk assessments across all lines of product that the Agency regulates. Following is an excerpt of main guidances promulgated by U.S. FDA for drug/biological and medical device products.

Drug and Biological Products

FDA's safety assessment of medicines does not diminish after drugs are approved for marketing. Although the pre-market phase of study is very intensive, much work remains to monitor approved drugs over time. No drug is risk-free, and it is not uncommon for new information to be discovered after a drug is on the market and being used by larger numbers of patients. Such information helps provide a better picture of drug risks, enables FDA to give healthcare professionals and patients the latest information on potential or newly identified risks, and strengthens FDA's ability to safeguard patients against unacceptable risks. On June 12, 2002, Congress reauthorized, for the second time, the Prescription Drug User Fee Act (PDUFA III). In the context of PDUFA III, FDA agreed to satisfy certain performance goals. One of those goals was to produce guidance for industry on risk management activities for drug and biological products. FDA published the following three final guidances in March 2005:

1. *Premarketing Risk Assessment* (Pre-marketing guidance)
2. *Development and Use of Risk Minimization Action Plans* (RiskMAP Guidance)
3. *Good Pharmacovigilance Practices and Pharmacoepidemiologic Assessment* (Pharmacovigilance guidance)

Each of the three final guidance documents focuses on one aspect of risk management. The pre-marketing guidance and the pharmacovigilance guidance focus on pre-marketing and post-marketing risk assessment, respectively. The RiskMAP guidance focuses on risk minimization. Together, risk assessment and risk minimization form what FDA calls risk management. Specifically, risk management is an iterative process of:

1. Assessing a product's benefit–risk balance
2. Developing and implementing tools to minimize its risks while preserving its benefits
3. Evaluating tool effectiveness and reassessing the benefit–risk balance
4. Making adjustments, as appropriate, to the risk minimization tools to further improve the benefit–risk balance

This four-part process should be continuous throughout a product's life cycle, with the results of risk assessment informing the sponsor's decisions regarding risk minimization.

Risk management is an iterative process designed to optimize the benefit–risk balance for regulated products. Risk assessment consists of identifying and characterizing the nature, frequency, and severity of the risks associated with the use of a product. Risk assessment occurs throughout a product's life cycle, from the early identification of a potential product, through the pre-marketing development process, and after approval during marketing.

Pre-marketing risk assessment represents the first step in this process, and this guidance focuses on risk assessment prior to marketing. It is critical to FDA's decision on product approval that a product's underlying risks and benefits be adequately assessed during the pre-marketing period. Sponsors seeking approval must provide from the clinical trials a body of evidence that adequately characterizes the product's safety profile. This guidance provides general recommendations for assessing risk. The adequacy of the assessment of risk is a matter of both quantity (ensuring that enough patients are studied) and quality (the appropriateness of the assessments performed, the appropriateness and breadth of the patient populations studied, and how results are analyzed). This guidance also addresses the qualitative aspects of risk assessment.

Although risk assessment continues through all stages of product development, this guidance focuses on risk assessment during the later stages of clinical development, particularly during phase 3 studies. The guidance is not intended to cover basic aspects of preclinical safety assessments (for example animal toxicity testing) or routine clinical pharmacology programs. Good clinical risk assessment in the later stages of drug development should be guided by the results of comprehensive preclinical safety assessments and a rigorous, thoughtful clinical pharmacology program (including elucidation of metabolic pathways, identification of possible drug–drug interactions, and determination of any effects from hepatic and/or renal impairment). These issues are addressed in other FDA guidances, as well as guidances developed under the auspices of the ICH.

Risk assessment during product development should be conducted in a thorough and rigorous manner; however, it is impossible to identify all safety concerns during clinical trials. Once a product is marketed, there is generally a large increase in the number of patients exposed, including those with comorbid conditions and those being treated with concomitant medical products. Therefore, post-marketing safety data collection and risk assessment based on observational data are critical for evaluating and characterizing a product's risk profile and for making informed decisions on risk minimization.

The pharmacovigilance guidance focuses on pharmacovigilance activities in the post-approval period. This guidance uses the term *pharmacovigilance* to mean all scientific and data-gathering activities relating to the detection, assessment, and understanding of adverse events. This includes the use of pharmacoepidemiologic studies. These activities are undertaken with the goal of identifying adverse events and understanding, to the extent possible, their nature, frequency, and potential risk factors.

Pharmacovigilance principally involves the identification and evaluation of safety signals. In this guidance document, *safety signal* refers to a concern about an excess of adverse events compared to what would be expected to be associated with a product's use. Signals can arise from post-marketing data and other sources, such as preclinical data and events associated with other products in the same pharmacologic class. It is possible that even a single well-documented case report can be viewed as a signal, particularly if the report describes a positive rechallenge or if the event is extremely rare in the absence of drug use. Signals generally indicate the need for further investigation, which may or may not lead to the conclusion that the product caused the event. After a signal is identified, it should be further assessed to determine whether it represents a potential safety risk and whether other action should be taken.

Medical Devices

FDA recognizes that there are risks associated with every medical device on the market. At the time of device approval, certain safety and effectiveness questions may not be fully resolved due to significant obstacles, such as the time and cost involved to address possible rare adverse events or long-term safety issues, and because controlled clinical studies do not fully represent the benefit–risk profile of a device when used in real-world clinical practice. Getting the right balance between pre-market and post-market data collection (specifically, where appropriate, a greater reliance on post-market collection) can reduce the extent of pre-market data submission and directly impact when patients will have access to high-quality, safe, and effective medical devices. But, greater reliance on post-market data collection could undermine patient safety if the necessary and timely data collection does not occur.

FDA has long applied post-market controls as a way to reduce pre-market data collection, where appropriate, while assuring that the statutory standard for approval of reasonable assurance of safety and effectiveness is still met.

An ample discussion on this topic is found on FDA's guidance *Balancing Premarket and Postmarket Data Collection for Devices Subject to Premarket Approval* issued on April 13, 2015. This guidance provides a resource for industry and FDA staff on how FDA determines when it is appropriate for manufacturers (mainly, sponsors of a PMA) to collect some data (clinical or nonclinical) in the post-market setting, rather than pre-market.

FDA believes that applying post-market controls to reduce pre-market data collection, when appropriate, improves patient access to safe and effective medical devices that are important to public health. As discussed in this guidance, there are certain circumstances where FDA may consider it acceptable to collect certain data in the post-market setting, rather than pre-market. FDA applies careful post-market monitoring to support the continued safety and effectiveness of the device.

Post-market data collection is a factor FDA considers as a part of making benefit–risk determinations. FDA may consider the collection of post-market data as a way to clarify the magnitude and effect of mitigations or as a way to develop additional information regarding benefits or risks for certain device types or in specific patient populations when making a benefit–risk determination. As part of FDA's benefit–risk determination, one of the factors that FDA considers is the degree of certainty of the probable benefits and probable risks of a device in the Agency's review of a PMA application. FDA states that:

> There is never 100% certainty when determining reasonable assurance of safety and effectiveness of a device or any other medical product. However, the degree of certainty of the benefits and risks of a device is a factor FDA considers when making benefit–risk determinations.

FDA recognizes that medical device approvals are not made with absolute certainty due to significant obstacles, such as the time and cost involved to address possible rare adverse events or long-term safety, and because clinical studies do not fully represent how a device will be used in clinical practice.

When making a determination of whether it is appropriate to collect certain data in the post-market setting, rather than pre-market, FDA considers, among other factors, the device's potential impact on public health. FDA may approve a device with a greater degree of uncertainty regarding the benefits and risks of the device if this uncertainty is sufficiently balanced by other factors, including the probable benefits of the device and the extent of post-market controls.

As discussed in the following, during the review of a PMA, FDA may consider it acceptable to collect certain data in the post-market setting, rather than pre-market, under certain circumstances when FDA has uncertainty regarding certain benefits or risks of the device but less uncertainty about the overall benefit–risk profile of the device at the time of pre-market approval. For example, uncommon or minor risks raised prior to PMA approval may, in appropriate instances, be addressed post-market; assessing the long-term benefit–risk profile of the device may also be deferred to the post-market phase. In addition, in cases where FDA has robust experience with the device type—typically a more mature technology—there may be less uncertainty about probable risks and benefits of the device, and FDA may leverage such experience to defer some data collection to the post-market setting.

However, many concerns have been raised about lowering pre-market standards and emphasizing post-market data even though the draft guidance admits that "Greater reliance on post-market data collection could undermine patient safety if the necessary and timely data collection does not occur."

Endnotes

1. For a list of currently approved REMS, see https://www.accessdata.fda.gov/scripts/cder/rems/index.cfm.
2. For in-depth coverage of this topic, see Rodriguez-Perez (2022).

6

Methodologies and Tools

Quality risk management supports a scientific approach to decision-making. It provides documented and reproducible methods for accomplishing steps of the quality risk management process based on current knowledge about assessing the severity, probability, and detectability of the risk. Traditionally, the medical product industry and regulators have been assessing and managing risk in a variety of informal ways based on, for example, compilation of observations, trends, and other information. Such approaches continue to provide useful information that might support topics such as handling of complaints, quality defects, deviations, and allocation of resources.

Quality risk management methods and the supporting statistical tools can be used in combination (for example, probabilistic risk assessment). Combined use provides flexibility that can facilitate the application of quality risk management principles. These techniques can be complementary, and it may be necessary to use more than one of them. The basic principle is that the chain of events is analyzed *step by step*. The degree of rigor and formality of quality risk management should reflect available knowledge and be commensurate with the complexity and/or criticality of the issue under evaluation.

SELECTION AND TYPES OF RISK ASSESSMENT TECHNIQUES

Risk assessment may be undertaken in varying degrees of depth and detail and using one or several methods ranging from simple to complex. The form of assessment and its output should be consistent with the risk criteria developed as part of establishing the context. In general terms, a suitable technique should have the following characteristics:

- It should be justifiable and commensurate to the situation under consideration.
- It should provide results in a form that enhances understanding of the nature of the risk and how it can be treated.
- It should be capable of being used in a manner that is traceable, repeatable, and verifiable.

Once the decision has been made to perform a risk assessment, and the objectives and scope have been defined, the techniques should be selected based on applicable factors, such as:

- Objectives of the study.
- Type and range of risks being analyzed.
- The potential magnitude of the consequences. The decision on the depth to which risk assessment is carried out should reflect the initial perception of consequences (although this may have to be modified once a preliminary evaluation has been completed).
- Degree of expertise and resources needed. A simple method well done may provide better results than a more sophisticated procedure poorly done as long as it meets the objectives and scope of the assessment. The effort put into the assessment should be consistent with the potential level of risk being analyzed.
- Regulatory or contractual requirements.

Various factors influence the selection of an approach to risk assessment, such as the availability of resources, the nature and degree of uncertainty in the data and information available, and the complexity of the application. Information regarding those factors is included later in this chapter under each specific technique.

Availability of Resources

Several factors related to available resources and capabilities that may affect the choice of risk assessment techniques include:

- The skills, experience, capacity, and capability of the risk assessment team
- Constraints on time and other resources within the organization
- The budget available if external resources are required

The Nature and Degree of Uncertainty

The nature and degree of uncertainty requires an understanding of the quality, quantity, and integrity of information available concerning the risk under consideration. This includes the extent to which sufficient information is available regarding the risk, its sources and causes, and its consequences. Uncertainty can stem from poor data quality or the lack of essential and reliable data. Uncertainty can also be inherent in the external and internal context of the organization.

Available data do not always provide a reliable basis for the prediction of the future. For unique types of risks or for new product types, historical data may not be available or there may be different interpretations of available data by different stakeholders (industry and regulators, for example). Those undertaking risk assessment need to understand the type and nature of the uncertainty and appreciate the implications for the reliability of the risk assessment results. These should always be communicated to decision-makers.

Complexity

Risks can be complex in themselves, as, for example, in complex systems that need to have their risks assessed across the system rather than treating each component separately and ignoring interactions. In other cases, treating a single risk can have implications elsewhere and can impact other activities. Consequential impacts and risk dependencies need to be understood to ensure that in managing one risk an intolerable situation is not created elsewhere. Understanding the complexity of a single risk or of a portfolio of risks of an organization is crucial for the selection of the appropriate method or techniques for risk assessment.

APPLICATION OF RISK ASSESSMENT DURING LIFE-CYCLE PHASES

Many activities, projects, and products can be considered to have a life cycle starting from initial concept and definition through realization to final completion, which might include removal of the medical product from the market. Risk assessment can be applied at all stages of the life cycle and is usually applied many times with different levels of detail to assist in the decisions that need to be made at each phase.

Life-cycle phases have different needs and require different techniques. For example, during the concept and definition phase, when an opportunity is identified, risk assessment may be used to decide whether to proceed or not. Where several options are available, risk assessment can be used to evaluate alternative concepts to help decide which provides the best balance of risks. During the design and development phase, risk assessment contributes to ensuring that system risks are tolerable and to identifying risks impacting subsequent life-cycle phases.

RISK ASSESSMENT TECHNIQUES AND TOOLS

Basic Risk Management Facilitation Methods

Some of the simple techniques that are commonly used to structure risk management (Tague 2005) by organizing data and facilitating decision-making are discussed next.

Flowcharts/Process Mapping. A flowchart is a picture that describes the steps of a process in sequential order. It is useful when trying to understand how a process is really being done. Unfortunately, many companies do not have flowcharts of their processes. When trying to solve a problem, it is a good idea to spend some time walking through the process and comparing how things are actually done (the real process) with how they should be done (the theoretical process required by applicable working instructions).

Flowcharts are easy-to-understand diagrams showing how steps in a process fit together. This makes them useful tools for communicating how processes work and for clearly documenting how a particular job is done. Furthermore, the act of mapping a process in flowchart format helps you clarify your understanding of the process and helps you think about where the process can be improved.

Flowcharts are simple diagrams that map out a process so it can easily be communicated to other people. To draw a flowchart, brainstorm the tasks and decisions made during a process and write them down in order. Then map these out in flowchart format using appropriate symbols for the start and end of a process, actions to be taken, and decisions to be made. Finally, challenge your flowchart to make sure it's an accurate representation of the process and that it represents the most efficient way of doing the job. This tool must be the first step in any risk assessment methodology.

Strengths
• Flowcharts are one of the best ways to understand a process. • They are also a good way of communicating the logic of a system to stakeholders. • Problems can be analyzed in a more effective way. • Flowcharts help in debugging processes.
Limitations
• Sometimes, a flowchart becomes very complex. • If alterations are required, the flowchart may require redrawing completely. • The essentials of what is done can easily be lost in the technical details of how it is done.

Brainstorming. The term "brainstorming" is often used very loosely to mean any type of group discussion. Brainstorming involves stimulating and encouraging free-flowing conversation among a group of knowledgeable people to identify potential failure modes and associated hazards, risks, criteria for decisions, and/or options for treatment. However, true brainstorming involves particular techniques to try to ensure that people's imaginations are triggered by the thoughts and statements of others in the group.

Effective facilitation is very important in this technique and includes stimulation of the discussion at kickoff, periodic guiding of the group into other relevant areas, and capture of the issues arising from the discussion. Brainstorming can be used in conjunction with other risk assessment methods (for example cause-and-effect diagram, fault tree analysis, and so on) described below or may stand alone as a technique for encouraging imaginative thinking at any stage of the risk management process and any stage of the life cycle of a system. Brainstorming places a heavy emphasis on imagination. It is therefore particularly useful when identifying risks of new technology, where there are no data, or where novel solutions to problems are needed.

Strengths
• Encourages imagination, which helps identify new risks and novel solutions. • Involves key stakeholders and hence aids communication overall. • Relatively quick and easy to set up.

Limitations
• Participants may lack the skill and knowledge to be effective contributors.
• Since it is relatively unstructured, it is difficult to demonstrate that the process has been comprehensive, for example, that all potential risks have been identified.
• There may be particular group dynamics where some people with valuable ideas stay quiet while others dominate the discussion. This can be overcome by anonymous computer brainstorming, using a chat forum, or nominal group technique.

Checklists. Checklists are lists of hazards, risks, or control failures that have usually been developed from experience, either as a result of a previous risk assessment or of past failures. A checklist can be used to identify hazards and risks or to assess the effectiveness of controls. They can be used at any stage in the life cycle of a product, process, or system. They may be used as part of other risk assessment techniques, but they are most useful when applied to check that everything has been covered after a more imaginative technique that identifies new problems has been applied.

Strengths
• May be used by nonexperts.
• If well designed, they combine wide-ranging expertise with an easy-to-use system.
• Can help ensure that common problems are not forgotten.
Limitations
• Tend to inhibit imagination in the identification of risks.
• Address the "known" part of the situation, not the "unknown."
• Encourage "tick the box" type behavior.
• Tend to be observation based, so miss problems that are not readily seen.

Delphi Method. The Delphi method (Linstone and Turoff 1975) is a structured communication technique, originally developed as a systematic, interactive forecasting method that relies on a panel of experts. Although the term is often now broadly used to mean any form of brainstorming, an essential feature of the Delphi method, as originally formulated, was that experts expressed their opinions individually and anonymously while having access to the other experts' views as the process progresses. The name "Delphi" derives from the Oracle of Delphi. The authors of the method were not happy with this name, because it implies "something oracular, something smacking a little of the occult." The Delphi method is based on the assumption that group judgments are more valid than individual judgments.

In the standard version, the experts answer questionnaires in two or more rounds. After each round, a facilitator provides an anonymous summary of the experts' forecasts from the previous round as well as the reasons they provided for their judgments. Thus, experts are encouraged to revise their earlier answers in light of the replies of other members of their panel. It is believed that during this process the range of the answers will decrease and the group will converge toward the "correct" answer. Finally, the process is stopped after a predefined criterion (for example, the number of rounds, achievement of consensus, stability of results), and the mean or median scores of the final rounds determine the results.

The Delphi method is based on the principle that forecasts (or decisions) from a structured group of individuals are more accurate than those from unstructured groups. This has been described using the term "collective intelligence." The Delphi method has been widely used for business forecasting and has certain advantages over another structured forecasting approach, prediction markets. The Delphi method was developed at the beginning of the Cold War to forecast the impact of technology on warfare. In 1944, General Henry H. Arnold ordered the creation of a report for the U.S. Army Air Corps on the future technological capabilities that might be used by the military. Different approaches were tried, but the shortcomings of traditional forecasting methods—such as the theoretical approach, quantitative models, or trend extrapolation—in areas where precise scientific laws have not been established yet quickly became apparent. The Delphi method was developed during the 1950s and 1960s by Olaf Helmer, Norman Dalkey, and Nicholas Rescher. Experts were asked to give their opinion on the probability, frequency, and intensity of possible enemy attacks. Other experts could anonymously give feedback. This process was repeated several times until a consensus emerged.

Strengths
• As views are anonymous, unpopular opinions are more likely to be expressed. • All views have equal weight, which avoids the problem of dominating personalities. • Achieves ownership of outcomes. • People do not need to be brought together in one place at one time.
Limitations
• It is labor-intensive and time-consuming. • Participants need to be able to express themselves clearly in writing.

Failure Mode and Effects Analysis (FMEA)

Within the medical product manufacturing setting, few tools are more appropriate than the failure mode and effects analysis (FMEA) (International Electrotechnical Commission 2006). In one of its guidances for industry (Q9(R1) Quality Risk Management), FDA summarizes some of the most common risk management tools. Among those tools, they include the FMEA establishing that:

> FMEA provides for an evaluation of potential failure modes for processes and their likely effect on outcomes and/or product performance. Once failure modes are established, risk reduction can be used to eliminate, contain, reduce, or control the potential failures. FMEA relies on product and process understanding. FMEA methodically breaks down the analysis of complex processes into manageable steps. It is a powerful tool for summarizing the important modes of failure, factors causing these failures, and the likely effects of these failures. (Food and Drug Administration 2023a)

Furthermore, it mentions that FMEA can be used to prioritize risks and monitor the effectiveness of risk control activities. Besides its definition and areas of application, the guide does not provide a methodology to carry out an effective FMEA. For that reason, a systematic approach is needed. That approach will consist of three interrelated tools: the process map, the cause-and-effect matrix (prioritization matrix), and the FMEA.

FMEA is an analysis technique that facilitates the identification of potential problems in the design or process by examining the effects of lower-level failures. Recommended actions or mitigation provisions are made to reduce the likelihood of the problem occurring and mitigate the risk if, in fact, it does occur.

FMEA or FMECA? The FMEA determines the effect of each failure mode and identifies single failure points that are critical. It may also rank each failure according to the criticality of a failure effect and its probability of occurring. Therefore, the failure mode, effects, and criticality analysis (FMECA) is the result of two steps:

- Failure mode and effects analysis (FMEA)
- Criticality analysis (CA)

FMECA is just FMEA with criticality analysis, and they are often used interchangeably. From now on, when I use the acronym FMEA I am referring to both FMEA and FMECA. There are many different flavors of FMEAs. They are inductive techniques using the question, "What happens if . . . ?" Elements of the process (for example, components) are analyzed one at a time, thus generally looking at a single-fault condition. This is done in a "bottom-up" mode, for example, following the procedure to the next higher functional system level.

FMEA provides for an evaluation of potential failure modes of processes and their likely effect on outcomes and/or product performance. Once failure modes are established, risk reduction can be used to eliminate, contain, reduce, or control the potential failures, thus creating preventive actions to minimize risks.

FMEA relies on product and process understanding. It methodically breaks down the analysis of complex processes into manageable steps. It is a powerful tool for summarizing the important modes of failure, factors causing these failures, and the likely effects of these failures. FMEA can be used to prioritize risks and monitor the effectiveness of risk control activities. FMEA can be applied to equipment and facilities and might be used to analyze a manufacturing operation and its effect on product or process. It identifies elements/operations within the system that render it vulnerable. The output/results of FMEA can be used as a basis for design or further analysis, or to guide resource deployment.

FMEA application in the medical product industry can be utilized for failures and risks associated with manufacturing processes; however, it is not limited to this application. The output of an FMEA is a relative risk "score" for each failure mode, which is used to rank the modes on a relative risk basis. FMEA can provide input into other analysis techniques such as fault tree analysis at either a qualitative or quantitative level.

Strengths
• Provides a basis for identifying potential root causes and developing effective preventive actions.
• Identifies reliability/safety-critical components. It is widely applicable to human, equipment, and system failure modes and to hardware, software, and procedures.
• Provides a foundation for other maintainability, safety, testability, and logistics analyses.
• Identifies component failure modes, their causes, and their effects on the system and presents them in an easily readable format.
• Avoids the need for costly equipment modifications in service by identifying problems early in the design process.
• Identifies single-point failure modes and requirements for redundancy or safety systems. Provides input to the development monitoring programs by highlighting key features to be monitored.
Limitations
• It is difficult to deal with redundancies and the incorporation of repair or preventive maintenance actions, as well as its restriction to single-fault conditions.
• It can only be used to identify single failure modes, not combinations of failure modes.
• FMEA/FMECA studies can be time-consuming and costly.
• FMEA/FMECA studies can be difficult and tedious for complex, multi-layered systems.

The FMEA Process. FMEAs need information about the elements of the system in sufficient detail to conduct meaningful analysis of the ways in which each element can fail. Information may include:

- A flowchart of the process or system being analyzed and its components
- An understanding of the function of each step of the process or component of the system
- Details of environmental and other operational parameters that may affect operation
- An understanding of the results of particular failures
- Historical data on failures, including failure rate data (if available), which will allow determination of the probability of failure

There are several ways this may be done. The most common method used in the regulated industry to classify each of the identified failure modes according to its criticality is the risk priority number (RPN), which is a semiquantitative measure of criticality obtained by multiplying numbers from rating scales (usually between 1 and 10) for consequence (severity) of failure, likelihood of occurrence of failure, and ability to detect the problem. A failure is given a higher priority if it is difficult to detect. Once potential failure modes and mechanisms are identified, preventive actions can be defined and implemented for the more significant ones.

The FMEA process is depicted in Figure 6.1.

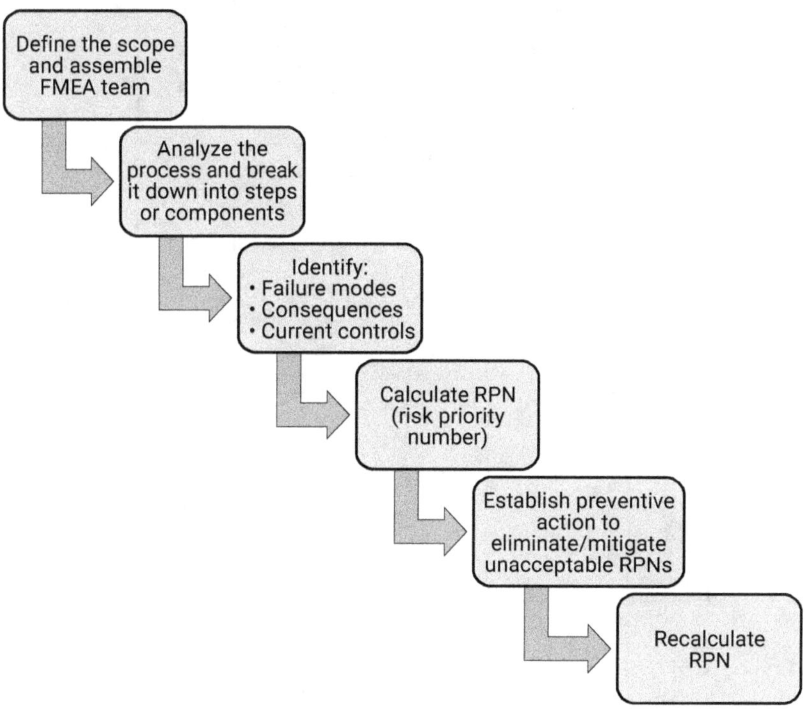

Figure 6.1 FMEA process.

The primary output of FMEA/FMECA is a list of potential failure modes—the failure mechanisms and effects for each component or step of a system or process (which may include information on the likelihood of failure). Information is also given on the causes of failure and the consequences to the system as a whole, and the detectability of the failure mode. See Figure 6.2 for an example of FMEA output.

Process step/input	Potential failure mode	Potential failure effects	SEV	Potential causes	OCC	Current controls	DET	RPN	Actions recommended	Resp.	Actions taken	SEV	OCC	DET	RPN
What is the process step/input under investi-gation?	In what ways does the key input go wrong?	What is the impact on the key output variables (customer requirements) or internal requirements?	How severe is the effect to the customer?	What causes the key input to go wrong?	How often does cause or FM occur?	What are the existing controls and procedures (inspection and test) that prevent either the cause or the failure mode? *Should include an SOP number.*	How well can you detect cause or FM?		What are the actions for reducing the occurrence of the cause, or improving detection?	Who's responsible for the recom-mended action?	What are the completed actions taken with the recalculated RPN? *Be sure to include completion month/year.*				
Filling/ feeder	Excessive vibration	Unfilled cavities	10	Lack of adequate preventive maintenance	8	Automatic inspection	7	560	Include the machine in the PM program	M. Peña	Machine was included in monthly PM program	10	2	3	60
Filling/ feeder	Excessive vibration	Cavities with more than one tablet	10	Lack of adequate preventive maintenance	8	Automatic inspection	7	560	Include the machine in the PM program	M. Peña	Machine was included in monthly PM program				0

Figure 6.2 Example of FMEA for tablet packaging.

Appendix C contains a case study using FMEA. Chapter 8 discusses some pitfalls when using FMEA as the only risk management tool.

Fault Tree Analysis

Fault tree analysis (FTA) is a type of analysis in which a failure is analyzed using Boolean logic (and/or) to combine a series of lower-level events (causal factors) until you reach their root causes (Tague 2005). This analysis method was originally developed to quantitatively determine the probability of a safety hazard in the field of safety engineering. FTA provides a method of breaking down chains of failures. A key addition permits the identification of combinations or interactions of events that cause other failure events (see example in Figure 6.3). There are two types of interaction:

a. *Several* items must fail together to cause another item to fail ("AND" combination)

b. Only *one* of a number of possible events must happen to cause another item to fail ("OR" combination)

The AND and OR symbols are called *gates*. They prevent the failure event above them from occurring unless their specific conditions are met. When several factors must happen simultaneously (AND relationship), you can avoid the failure simply by controlling one of them (the easiest or the cheapest). When any of several causal factors can create the failure, then you must fix all of them.

The tree is constructed working backward from a known event or failure and asking why it happened. The answer will represent the factor that directly caused the failure. Continuing with the why questioning will allow us to reach fundamental events or root causes. In other words, FTA is a very good tool to help us understand how an event occurred. It is best used when working with complex issues with several interrelated causes of failure.

FTA is a deductive, top-down approach to failure mode analysis aimed at analyzing the possible causes of an undesired event or failure. This contrasts with FMEA, an inductive, bottom-up analysis method aimed at analyzing the effects of single-component or single-function failures on equipment or systems. In terms of the CAPA system, you can define FTA as a reactive investigation tool (the failure already happened). FMEA should, ideally, be used proactively (during the design phase of a process) to anticipate failure modes and generate preventive actions.

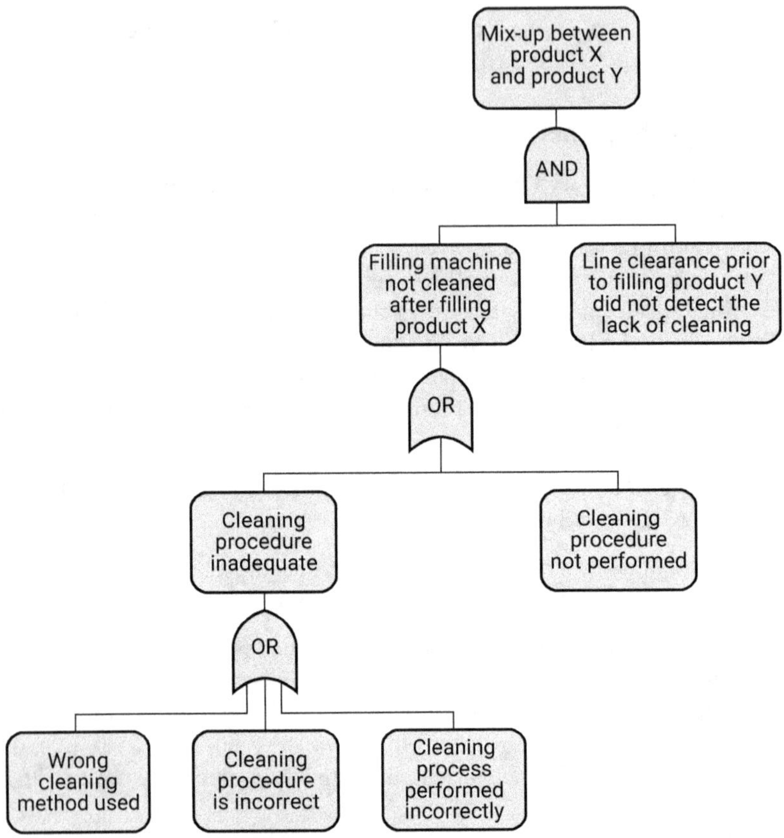

Figure 6.3 Fault tree analysis example.

FTA is a technique for identifying and analyzing factors that can contribute to a specified undesired event (called the "top event"). In a deductive manner, starting with the top event, the possible causes or fault modes of the next lower functional system level causing the undesired consequence are identified. The factors identified in the tree can be events that are associated with component hardware failures, human errors, or any other pertinent events that lead to the undesired event.

FTA can be used to establish the pathway to the root cause of the failure. FTA can be used to investigate complaints or deviations in order to fully understand their root cause and ensure that intended improvements will fully resolve the issue and not lead to other issues (solve one problem yet cause a different problem). FTA is an effective tool for evaluating how multiple factors affect a given issue. The output of an FTA includes a visual representation of failure modes. It is useful both for risk assessment and in developing monitoring programs.

Strengths
• It is a disciplined approach that is highly systematic but at the same time sufficiently flexible to allow analysis of a variety of factors, including human interactions. The "top-down" approach focuses attention on those effects of failure that are directly related to the top event.
• FTA is especially useful for analyzing systems with many interfaces and interactions.
• The pictorial representation leads to an easy understanding of the system behavior and the factors included, but as the trees are often large, processing of fault trees may require computer systems.
• Logic analysis of fault trees is useful in identifying simple failure pathways in a very complex system where particular combinations of events that lead to the top event could be overlooked.
Limitations
• Uncertainties in the probabilities of base events are included in calculations of the probability of the top event. This can result in high levels of uncertainty where base event failure probabilities are not known accurately; however, a high degree of confidence is possible in a well-understood system.
• In some situations, causal events are not bound together, and it can be difficult to ascertain whether all important pathways to the top event are included.
• The fault tree is a static model; time interdependencies are not addressed.
• Fault trees can only deal with binary states (failed/not failed).
• While human error modes can be included in a qualitative fault tree, in general, failures of degree or quality, which often characterize human error, cannot easily be included.
• A fault tree does not enable domino effects or conditional failures to be included easily.

Hazard Analysis and Critical Control Points (HACCP)

HACCP is a systematic, proactive, and preventive tool for assuring product quality, reliability, and safety (World Health Organization 2003). It is a structured approach that applies technical and scientific principles to analyze, evaluate, prevent, and control the risk or adverse consequences of hazards due to the design, development, production, and use of products.

Traditionally, the HACCP methodology has been considered to be a food safety management system. It aims to prevent known hazards and to reduce the risk that they will occur at specific points in the food chain. The same principles are also increasingly being applied in other industries, such as the auto, aviation, chemical, and medical product industries. Hazards affecting quality are controlled to a certain extent through the validation of critical operations and processes in the manufacture of medical products in accordance with applicable cGMP. However, cGMP does not cover the safety of the personnel engaged in manufacture, while both aspects are covered by HACCP. Safety hazards are common in the manufacture of active pharmaceutical ingredients and will require special precaution and control measures.

Procedures, including cGMP, address operational conditions and provide the basis for HACCP. HACCP is a systematic method for the identification, assessment, and control of safety hazards. Such hazards are defined as biological, chemical, or physical agents or operations that are reasonably likely to cause illness or injury if not controlled. In the manufacture of pharmaceuticals, these may include the manufacture of certain antibiotics, hormones, cytotoxic (carcinogenic) substances, or other highly active pharmaceuticals, together with operations such as fluid-bed drying and granulation, which are examples of hazard unit operations in the pharmaceutical manufacturing arena. The use of inflammable solvents (solutions) and certain laboratory operations may also constitute hazards.

HACCP might be used to identify and manage risks associated with physical, chemical, and biological hazards (including microbiological contamination). HACCP is most useful when product and process understanding is sufficiently comprehensive to support identification of critical control points. The output of a HACCP analysis is risk management information that facilitates monitoring of critical points not only in the manufacturing process but also in other life-cycle phases. The intent of HACCP is to help prevent known hazards and to reduce the risk that they will occur at any point in a process through the execution of seven core principles, which are described in Figure 6.4.

The heart of an effective HACCP system focuses on the continuing control and monitoring of the identified hazards (principles 2, 3, and 4). A manufacturer demonstrates the effectiveness of established control measures (principles 5 and 6) by establishing methodically documented process mapping, process hazard analysis, and a critical control plan (principle 7).

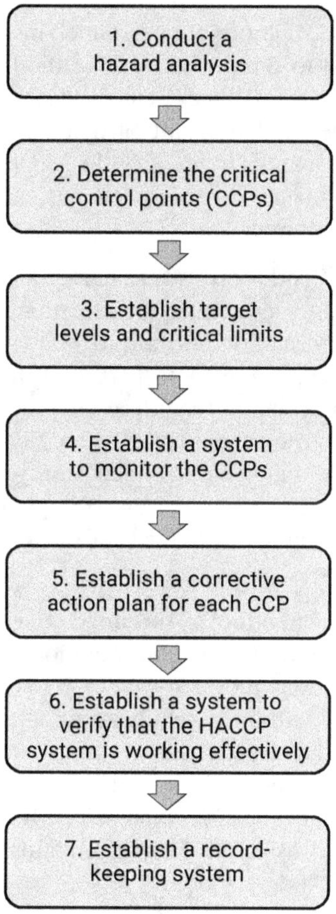

Figure 6.4 HACCP principles.

Developing a HACCP Plan. There are a total of 12 tasks required in developing a HACCP plan, and these are designed to ensure that the seven principles are applied correctly. Principle 1, which is to conduct a hazard analysis, requires that the first five tasks have all been addressed in a logical and honest manner so all real hazards associated with the commodity have been identified. The 12 tasks are discussed briefly below.

Task 1—Establish a HACCP Team. To fully understand the commodity system and be able to identify all likely hazards and CCPs, it is important that the HACCP team is made up of people from a wide range of disciplines.

The team should include appropriate personnel, including a team leader to convene the group and to direct the work of the team, ensuring that the concept is properly applied. This person must be familiar with the technique, be a good listener, and allow all participants to contribute. The first activity of the HACCP team is to identify the scope of the study. This will make the task more manageable, and specialists can be added to the team as and when they are required.

Task 2—Describe the Product. To start a hazard analysis, a full description of the product, including customer specifications, should be prepared. This should include information relevant to safety, regulation/target level, composition, and physical/chemical properties of the raw materials and the final product. Also, information regarding how the product is to be packaged, stored, and transported should also be considered together with facts regarding its shelf life and recommended storage temperatures. Where appropriate, labeling information and an example of the label should be included. This information will help the HACCP team identify real hazards associated with the process.

Task 3—Identify the Product's Intended Use. How the product is intended to be used is an important consideration. Information on whether the product will be consumed directly, or be cooked (in the case of food), or be further processed will all have a bearing on the hazard analysis (see task 6). The nature of the target group (customers/patients) for the product may also be relevant, particularly if it includes susceptible groups such as infants, the elderly, or pregnant women. The likelihood of misuse of a product should also be considered. This information can be recorded on the same form as the product description.

Task 4—Draw Up the Process Flow Diagram. The first function of the team is to draw up a detailed flow diagram of the process or system, or that part of it that is relevant.

Task 5—On-Site Confirmation of Flow Diagram. Upon completion of the flow diagram, members of the team should visit the process (for example, warehouse or manufacturing area) to compare the information present on the flow diagram with what actually happens in practice. This is known as "walking the line," a step-by-step practice to check that all information regarding materials, practices, controls, and so on have been taken into consideration by the team during the preparation of the flow diagram.

Task 6—Identify and Analyze Hazard(s)—(Principle 1). Effective hazard identification and hazard analysis are the keys to a successful

HACCP plan. All real or potential hazards that may occur in each ingredient and at each stage of the commodity system should be considered. Safety hazards for HACCP programs have been classified into three types:

1. *Biological.* Typically, microbiological, viral, or parasitic contamination with pathogens.
2. *Chemical.* Cross-contamination situations due to raw material or finished product mix-ups are a common (and very dangerous) situation in medical product manufacturing.
3. *Physical.* Contaminants such as broken glass, hair, pieces of clothes or gloves, metal fragments, insects, or stones.

The probability that a hazard will occur is called a *risk*. The risk may take a value from zero to one depending on the degree of certainty that the hazard will be absent or that it will be present. After hazard identification, a hazard analysis must be conducted to understand the relative health risk to man or animal posed by the hazard. It is a way of organizing and analyzing the available scientific information on the nature and size of the health risk associated with the hazard. The risk may have to be assessed subjectively and simply classified as low, medium, or high. Only those hazards considered by the HACCP team to present an unacceptable risk of being present are taken forward to principle 2.

Once a safety hazard has been identified, appropriate control measures should be considered. These are any action or activity that can be used to control the identified hazard such that it is prevented, eliminated, or reduced to an acceptable level.

Task 7—Determine the Critical Control Points (CCPs)—(Principle 2). Each step in the flow diagram, within the scope of the HACCP study, should be taken in turn, and the relevance of each identified hazard should be considered. It is also important to remember the stated scope of the HACCP analysis at this stage. The team must determine whether the hazard can occur at this step, and if so, whether control measures exist. If the hazard can be controlled adequately, is not best controlled at another step, and is essential for process safety, then this step is a CCP for the specified hazard. A decision tree can be used to determine CCPs. However, the HACCP team's judgment, expertise, and knowledge of the process are the major factors in establishing CCPs. If a step is identified where a safety hazard exists but no adequate control measures can be put in place either at this step or subsequently, then there is a high-risk situation that must be attended to immediately.

Task 8—Establish Critical Limits for Each CCP—(Principle 3). Critical limits must be specified and validated for each CCP. Criteria often used include measurements of process parameters such as product concentration. In the case of food safety, temperature, cooking time, moisture level, pH, water activity, and sensory parameters such as visual appearance are often used. All critical limits, and the associated permissible tolerances, must be documented in the HACCP plan worksheet and included as specifications in operating procedures and work instructions.

Task 9—Establish a Monitoring Procedure—(Principle 4). Monitoring is the best mechanism for confirming that critical limits at each CCP are being met. The method chosen for monitoring must be specific and have enough sensitivity to produce a meaningful result that is detectable with available monitoring equipment. This is imperative so corrective action can be taken as quickly as possible so loss of product will be avoided or minimized. Monitoring can be carried out by observation or by measurement of samples taken in accordance with a statistically based sampling plan. Monitoring by visual observation is basic but gives rapid results, and can therefore be acted on quickly.

Task 10—Establish Corrective Action—(Principle 5). If monitoring indicates that critical limits are not being met (the process is out of control) corrective action must be taken immediately. The corrective action should take into account the worst-case scenario but must also be based on the assessment of hazards, risk, and severity, and on the final use of the product. Corrective actions must ensure that the CCP has been brought back under control. They must also include appropriate disposition of any affected product. Whenever possible, an alarm system should be introduced that will activate when monitoring indicates that the critical limit is being approached. Preventive actions can then be applied to prevent the need for any product disposition.

Task 11—Verify the HACCP Plan—(Principle 6). Once the HACCP plan has been drawn up and all the CCPs have been validated, the complete plan must be verified. Once the HACCP plan is in routine operation, it must be verified and reviewed at regular intervals. The appropriateness of CCPs and control measures can thus be determined, and the extent and effectiveness of monitoring can be verified. Ways in which the system can be verified include:

- Collecting samples for analysis by a method different from the monitoring procedure (asking questions of staff, especially CCP monitors)
- Observing operations at CCPs (audit of the process)

Task 12—Keep Records—(Principle 7). Record keeping is an essential part of the HACCP process. It demonstrates that the correct procedures have been followed from the start to the end of the process, offering product traceability. It provides a record of compliance with the critical limits set and can be used to identify problem areas. Usually, full compliance with this task is not a problem for the highly regulated medical product industry.

Strengths
• HACCP is a highly structured process that provides documented evidence for quality control as well as identifying and reducing risks.
• Emphasis on control, not detection.
• Better risk control throughout the process rather than relying on final product inspection.
• Allows the identification of hazards introduced through human actions and how these can be controlled at the point of introduction or subsequently.
Limitations
• Requires comprehensive process understanding—not ideally suited for little-known hazards or processes.
• Does not quantify or prioritize risks.
• Does not quantify impact of additional controls on reducing risk.
• Appropriate controls need to be defined.

Hazard and Operability Analysis (HAZOP)

HAZOP is the acronym for *hazard and operability analysis* and is a structured and systematic examination of a planned or existing product, process, procedure, or system. It is a technique used to identify risks to people, equipment, environment, and/or organizational objectives. The HAZOP technique was initially developed to analyze chemical process systems, but it has since been extended to other types of systems and also to complex operations and to software systems.

HAZOP is similar to an FMEA. It is a systematic technique for identifying hazards and operability problems. While the use of HAZOP studies in the chemical industry focuses on deviations from design intent, there are alternative applications for a medical product developer.

A HAZOP can be applied to the operation/function of a medical device or to a process used in the manufacture or maintenance/service of medical products (for example, sterilization) that can have a significant impact on the function of a product. It can also be applied to manufacturing processes, including outsourced production and formulation, as well as the upstream suppliers, equipment, and facilities for pharmaceutical products. It has also been used primarily in the pharmaceutical industry for evaluating process safety hazards. Very similar to the HACCP, the output of a HAZOP analysis is a list of critical operations for risk management. This facilitates regular monitoring of critical points in the manufacturing process.

There are several types of HAZOP:

- *Process HAZOP.* Originally developed to assess plants and process systems
- *Human HAZOP.* Focused on human errors rather than technical failures
- *Procedure HAZOP.* Review of procedures or operational sequences
- *Software HAZOP.* Identification of possible errors in the development of software

HAZOP is a technique for identifying potential hazards in a system and identifying operability problems likely to lead to nonconforming products. HAZOP is based on a theory that assumes risk events are caused by deviations from design or operating intentions. This tool has a unique and distinctive feature, which is that such deviations are facilitated by using sets of "guide words" as a systematic list of deviation perspectives. It is generally carried out by a multidisciplinary team during a set of meetings where those guide words are applied, in turn, to all the process parameters in order to identify unexpected and yet credible deviations from the design/process intent. The assumption is that

$$\text{Guide word} + \text{Parameter} \rightarrow \text{Deviation}$$

Figure 6.5 presents a list of the most commonly used guide words.

HAZOP guide words are key supporting elements in the execution of a HAZOP analysis. According to IEC Standard 61882:

> The identification of deviations from the design intent is achieved by a questioning process using predetermined "guide words." The role of the guide word is to stimulate imaginative thinking, to focus the study, and elicit ideas and discussion.

Guide word	Meaning	Examples
No (not, none)	None of the design intent is achieved	• Preventive maintenance was not performed • No sterilization process was performed
More (more of, higher)	Quantitative increase in a parameter	• Overconcentrated drug product • Infusion pump delivering more dosage than needed
Less (less of, lower)	Quantitative decrease in a parameter	• Customer receiving less product than expected (lower pill count or fill volume) • Drug product underconcentrated
Other than (other)	Complete substitution—another activity takes place	• Product mix-ups • Product sterilized with an incorrect method
Part of	Only some of the design intention is achieved	• Device kit missing one component • No detergent used for cleaning, only water
As well as (more than)	An additional activity occurs	• An additional component was added to the formulation of the batch
Reverse	Logical opposite of the design intention occurs	• Sterile injectable drug with bacterial contamination
Early/late	Timing is different from the intention	• Detergent added too early (prior to pre-rinse of soiled equipment)
Before/after	Step (or part of it) performed out of order	• Components added in incorrect order to the batch

Figure 6.5 HAZOP guide words.

Strengths
• Involves a simple and intuitive tool within a multidisciplinary team environment.
• Generates solutions and risk treatment actions.
• Applicable to a wide range of systems, processes, and procedures.
• Allows consideration of the causes and consequences of human error.
Limitations
• Detailed analysis can be very time-consuming and therefore expensive.
• Discussion can be focused on detail issues of design, and not on wider or external issues.
• The process relies on the expertise of the designers, who may find it difficult to be sufficiently objective to seek problems in their designs.

Cause-and-Effect Analysis

Cause-and-effect analysis is a structured method for identifying possible causes of an undesirable event or problem (Rodríguez-Pérez 2022). It organizes the possible contributory factors into broad categories so all possible hypotheses can be considered. It does not, however, by itself point to the actual causes, since these can only be determined by real evidence and empirical testing of hypotheses. The information is organized in either a cause-and-effect diagram (also called a *fishbone diagram* or *Ishikawa diagram*) or sometimes a tree diagram. Cause-and-effect diagrams are used to analyze and find the root cause of a certain effect or problem. They are also known as fishbone diagrams because their shape is similar to a fish skeleton. Cause-and-effect diagrams are considered one of the seven basic tools of quality management.

The cause-and-effect diagram focuses on the causes rather than the effects. Because there may be a number of causes for a particular problem, this tool helps us to identify the root cause of the problem in an uncomplicated manner. This tool allows brainstorming in a structured format similar to an affinity diagram, where potential causes are grouped into logical categories such as materials, manpower, methods, machines, environment, and so on. Potential causes can be further tracked back to the root cause using the 5 Whys technique. Another way to do this is to examine the problem using typical categories known as the 5Ms and 1E:

- *Manpower.* Anyone involved with the process
- *Methods.* How the process is performed and the specific requirements for doing it, such as policies, procedures, instructions, and regulations
- *Machines.* Any equipment, computers, or tools required to accomplish the task
- *Materials.* Raw materials, parts, and components used to produce the final product
- *Measurements.* Data generated from the process that are used to evaluate it
- *Environment.* Environmental conditions such as humidity and temperature and the culture in which the process operates

Figure 6.6 depicts a typical fishbone diagram.

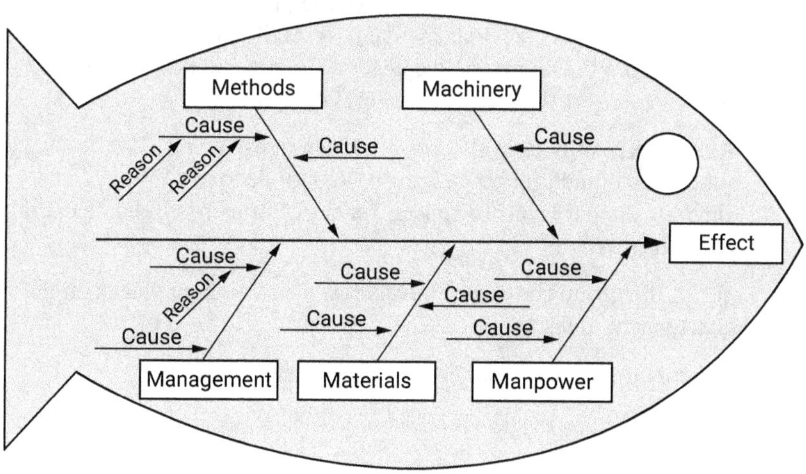

Figure 6.6 Cause-and-effect (fishbone) diagram example.

5 Whys

One of the most useful tools in science is *why*. As an investigator, you are searching for the true causes of your problems. If you don't know the cause, you cannot implement corrective action. Remember the definition of corrective action: action to avoid the recurrence of root causes. There are plenty of root cause analysis tools available to you, most of them more complicated to understand than your problem, but all of them are based on the word *why*.

The best way to find your way from the symptoms to the root causes is to ask *why* until the nature of your problem and its solution become clear. This tool (the 5 Whys) is attributed to Sakichi Toyoda, one of the founders of the Toyota Company.

Why five? It is postulated that five iterations is generally sufficient to get to a root cause. My recommendation is that you keep asking *why* until a *fixable* root cause is found.

The technique is very simple to use. Using a cross-functional teamwork environment:

- Agree on a problem statement (effect). Brainstorm the major categories of causes of the problem, or use the six generic headings.
- Brainstorm all the possible causes of the problem. Ask, "Why does this happen?" As each idea is given, the facilitator writes it as a branch from the appropriate category. Causes can be written in several places if they relate to several categories.
- Again ask, "Why does this happen?" about each cause. Write subcauses branching off the causes. Continue to ask "Why?" and generate deeper levels of causes. Layers of branches indicate causal relationships.
- When the group runs out of ideas, focus attention on places on the chart where ideas are few.

Figure 6.7 contains an example of the use of this tool.

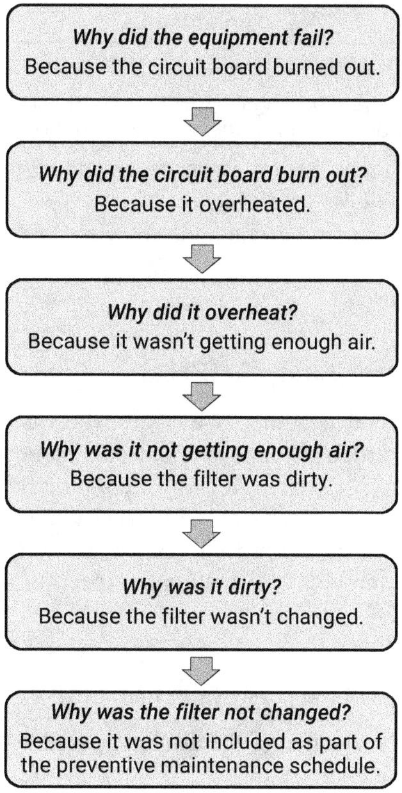

Figure 6.7 Example of use of the 5 Whys.

Strengths
• Involvement of applicable experts working in a team environment.
• Structured analysis.
• Consideration of all likely hypotheses.
• Graphical, easy-to-read illustration of results.
• Helps to identify areas where further data are needed.
• Can be used to identify contributory factors to wanted as well as unwanted effects.

Limitations
• The team may not have the necessary expertise.
• It is not a complete process in itself and needs to be a part of a root cause analysis to produce recommendations.
• It is a display technique for brainstorming rather than a separate analysis technique.
• Interactions between the categories may not be considered adequately.

Preliminary Hazard Analysis (PHA)

PHA is a simple, inductive method of analysis whose objective is to identify the hazards and hazardous situations and events that can cause harm for a given activity, facility, or system. It is most commonly carried out early in the development of a product or process, when there is little information on design details or operating procedures, and can often be a precursor to further studies or to provide information for specification of the design of a system.

The tool consists of:

1. Identification of the possibilities that the risk event happens
2. Qualitative evaluation of the extent of possible injury or damage to health that could result
3. A relative ranking of the hazard using a combination of severity and likelihood of occurrence
4. Identification of possible remedial measures

The results obtained can be presented in different ways such as tables and trees.

Strengths
• Can be used when there is limited information.
• Allows risks to be considered very early in the system life cycle.
Limitations
• Provides only preliminary information.
• It is not comprehensive; nor does it provide detailed information on risks and how they can best be prevented.

Risk Ranking and Filtering

Risk ranking and filtering is a tool for comparing and prioritizing (ranking) risks (see Food and Drug Administration 2004b). Risk ranking of complex systems typically involves evaluation of multiple diverse quantitative and qualitative factors for each risk. The tool involves breaking down a basic risk question into as many components as needed to capture factors involved in the risk. These factors are combined into a single relative risk score that can then be used for ranking risks. "Filters," in the form of weighting factors or cutoffs for risk scores, can be used to scale or fit the risk ranking to management or policy objectives.

Risk ranking (represented in Figure 6.8) and filtering (represented in Figure 6.9) can be used to prioritize manufacturing sites for inspection/audit by regulators or industry. Risk ranking methods are particularly helpful in situations in which the portfolio of risks and the underlying consequences to be managed are diverse and difficult to compare using a single tool. Risk ranking is useful for management in evaluating both quantitatively assessed and qualitatively assessed risks within the same organizational framework.

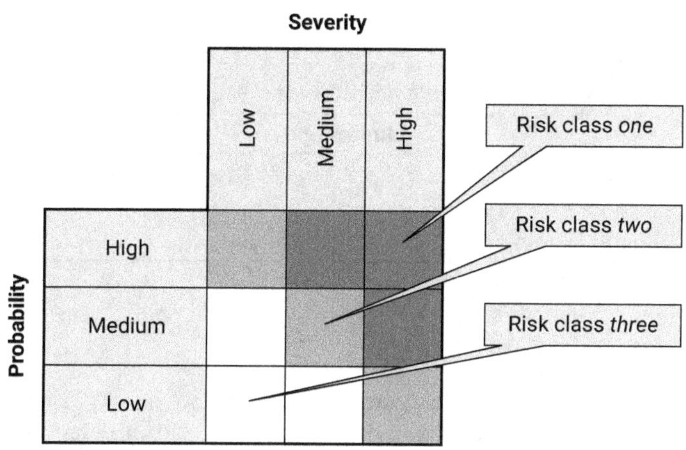

Figure 6.8 Risk ranking.

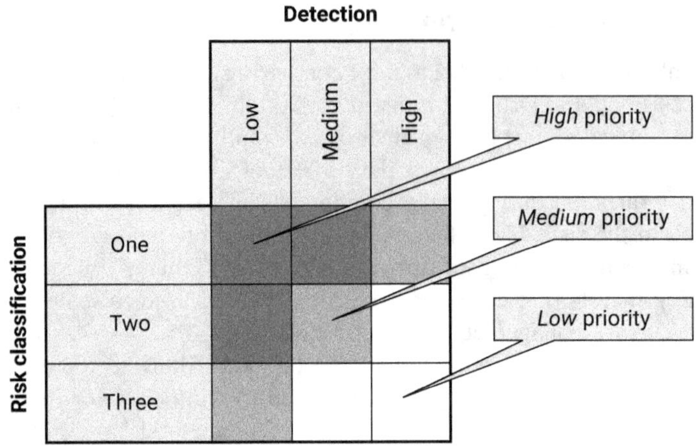

Figure 6.9 Risk filtering.

Strengths
• Can be used for highly complex issues.
• Very flexible.
• Scalable to include multiple risk factors.
Limitations
• Sometimes it is difficult to establish risk factors and adequate evaluation criteria.
• Difficult to correlate results with absolute risk.

Supporting Statistical Tools

Statistical tools can support and facilitate quantitative quality risk management. They can enable effective data assessment, aid in determining the significance of the data set(s), and facilitate more-reliable decision-making. The list of statistical tools commonly used in the FDA-regulated industry can be very extensive. Among the main tools are:

- Control charts
  - Acceptance control charts (see ISO 7966)
  - Control charts with arithmetic average and warning limits (see ISO 7873)

- Cumulative sum charts (see ISO 7871)
- Shewhart control charts (see ISO 8258)
- Weighted moving average

- Design of experiments (DOE)
- Histograms
- Pareto charts
- Process capability analysis

There are hundreds of very good statistical books. *Process Monitoring and Improvement Handbook,* second edition (2019) by Manuel E. Peña-Rodríguez, published by ASQ Quality Press, is an excellent reference for these topics.[1]

Endnote

1. See Peña-Rodriguez (2019).

7

Practical Applications of Quality Risk Management to the Life Science Manufacturing Industry

This chapter is intended to identify potential uses of quality risk management principles and tools by industry and regulators. However, the selection of risk management tools is completely dependent on specific facts and circumstances. These examples are provided for illustrative purposes and only suggest potential uses of quality risk management. Appendices A and C contain several examples of specific applications within the FDA-regulated environment.

MANAGEMENT

Risk management is a very important and critical process within a quality management system. Practically all major activities carried out by a medical product company must pass through the filter of risk management, from the prioritization of investigations when deviations occur, to the schedule and calendar for internal and supplier audit. It is difficult to imagine one area or department where the risk management process does not apply. Today, risk management is a standard and accepted quality system practice used by industry and regulators to facilitate good decision making about risk identification, resource prioritization, and risk mitigation or elimination, as appropriate. Even FDA is developing a modern risk-based medical product quality assessment system that focuses on critical product quality attributes and how they relate to the safety and effectiveness of those medical products.

Management of medical product companies must use risk management to review current interpretations and application of regulatory expectations. Following is a brief discussion of eight of these areas.

Internal Audit

Risk management tools can be used to define the frequency and scope of internal audits. The overall administration of the internal audit program is the responsibility of the quality unit. The quality unit is responsible for evaluating site operations and systems to identify the scope, frequency, and resources necessary to support an effective internal audit program as well as to determine the effectiveness of the conducted audits. It is a common practice to periodically audit every department (manufacturing, packaging, quality, training, warehouse, and so on) according to a predetermined schedule, and without using any of the many advantages that risk management/prioritization would provide.

For example, you can use risk ranking and filtering techniques to create a risk score for each department or quality system element under the internal audit program. I do recommend considering factors such as:

- Impact of the department/quality system element related to providing safe and effective products to your customer
- Compliance status and regulatory history of the department/element
- Results of previous audits/inspections
- Major changes since the last audit (for example, a new process introduced in the area, large turnover of personnel, and so on)
- CAPA system indicators (nonconformities, recalls, or complaints related to the area, and so on)

One element of auditing where risk management techniques are very often used is the classification of each audit finding and observation based on risk. Typically, the outcome of the audit and each individual finding/observation are classified using scores such as the following:

- Critical
- Major
- Minor

Training

Training is one of the most important elements of an effective and efficient quality system. Companies must provide adequate training resources. FDA regulations as well as international standards (for example ISO 13485:2016 or ICH Q10) refer to "competency" of those human resources in terms of education, training, inherent skills, and work experience.

Training provides skills and/or knowledge to adequately perform a job. Personnel should be qualified to do the operations that are assigned to them in accordance with the nature of, and potential risk of, their operational activities. On the other hand, managers should define appropriate qualifications for each position to help ensure that individuals are assigned appropriate responsibilities. Personnel should also understand the effect of their activities on the product and the customer. Job descriptions should include requirements such as scientific and technical knowledge, process and product knowledge, and/or risk assessment abilities to appropriately execute certain functions.

Continued training is critical to ensure that the employees remain proficient in their operational functions. Typical training should cover the policies, processes, procedures, and written instructions related to operational activities, the product/service, the quality system, and the desired work culture. Training should focus on both the employees' specific job functions and the related regulatory requirements. Managers are expected to establish training programs that include the following:

- Evaluation of training needs
- Provision of training to satisfy these needs
- Evaluation of the effectiveness of training
- Documentation of training and/or retraining

Risk assessment techniques must be used to determine the appropriateness of initial and/or ongoing training sessions based on the education, experience, and working habits of staff, as well as on a periodic assessment of previous training (for example, its effectiveness). Also, it is crucial to identify the training, experience, qualifications, and physical abilities that allow personnel to perform an operation reliably and with no adverse impact on the quality of the product. Therefore, the job description for each position within a medical product manufacturing plant must be developed using risk management techniques.

Even though training efforts are generally considered as part of the basic foundation of the quality system, cost reduction efforts, often tied to outsourcing activities, are hitting this critical area very hard. Many of the quality disasters related to medical products can be linked to poor or ineffective training efforts. Within this industry, training departments are being dismantled everywhere, or at least reduced to their minimal expression. Training dollars are seen as a cost instead of as an investment, and this is not just an offshoring problem. Local outsourcing activities within the United States and Western Europe are largely based only on cost reduction considerations, and training efforts are second to none during cost-cutting activities. An example of the use of risk analysis as part of the training program can be found in Falce (2015).

Error Risk Reduction[1]

According to a recent report published in the *British Medical Journal,* medical errors are the third leading cause of death in the United States, right after heart disease and cancer, causing at least 250,000 deaths every year, indicating that patient safety efforts fall far short. In other words, 10 percent of U.S. deaths are due to preventable medical mistakes (Makary and Daniel 2016).

For many years we considered human errors or mistakes the cause of a mishap or problem. Human error, under whatever label (procedures not followed, lack of attention, or simply error), was the conclusion of the investigation. Very often it was coupled with some kind of training activity (most frequently retraining) as corrective action. We even have an old adage—"To err is human"—to explain it. Human errors cannot be eliminated or even significantly reduced simply by telling operators to be more careful. This simplistic approach does not work because we are not addressing any root cause. Human error is more a symptom than a cause.

The aim of this chapter is to focus on the human side of quality and manufacturing errors. The way we look at the human side of these problems has evolved during the past few decades. Industrial psychologists and human reliability professionals took command during the investigation of catastrophic accidents (Chernobyl, Challenger, and aviation accidents), and the old conceptions changed. Now, we see human errors as the symptoms of deeper causes. In other words, human errors are consequences, not causes.

Most of the time, even though accidents are classified as mechanical or some other kind of failure, the roots go back to a human who made an error. For this discussion, human error excludes deliberate actions with harmful intent; these are considered sabotage. In the medical products regulated environment, the combination of human error and retraining is still indicated in the root cause and corrective action sections of many investigation reports.

Human errors are not a uniform collection of unwanted acts. Lack of attention plays a significant role in all categories of human error. Slips, lapses, and mistakes are all more common when situational factors (fatigue, workload, multitasking, and boredom) divert our attention. In the regulated industries, these factors should be negligible; you should not be relying on memory for procedures and instructions. Batch records and device history records exist for that exact purpose. You must read the instructions, execute the task, and then document the action performed.

U.S. FDA and foreign regulators require the consideration of human factors during the development of medical devices. In the landmark 1996 guidance *Do It By Design—An Introduction to Human Factors in Medical Devices,* FDA established design requirements to avoid user errors. More recently, the Agency introduced those concepts of human reliability to drug and biotechnology manufacturers.

Some statistics related to human error are:

- Ninety-nine percent of accidental losses (except for natural disasters) begin with a human error.
- The root causes of the vast majority of accidents are management system weaknesses.
- Eight percent of men are color-blind, while only one in 200 women have the condition.
- Eighty percent of medical product recalls due to incorrect expiration date or incorrect lot/batch number are caused by a transposition of digits.
- One and a half million Americans are injured every year by drug errors in hospitals, nursing homes, and doctor's offices (patients' own medication mix-ups are not included).
- On average, every hospitalized patient is subject to (at least) one medication error per day.
- Seventeen hours of work without a break is the same as being legally drunk.
- The worst time period for human errors is 2 a.m. to 5 a.m.
- About 15 percent of human errors are due to acquired habits.
- Human error accounts for 90 percent of road accidents.
- The rate of error and mistakes for most procedure-based tasks is one per hundred.
- Average workers are interrupted every 11 minutes and then spend almost one-third of their day recovering from these distractions.

Human factors is the common term given to the widely recognized discipline of addressing interactions in the work environment between people, the facility, and its management systems. Although individuals make errors and mistakes, it is usually because human factors issues were not properly addressed. In addition to being a key and critical aspect of process safety management, human factors are also the basis of sound process quality management. Human factors play an important role in process performance by using scientific knowledge and principles from many disciplines to reduce the frequency of errors and accidents. However, the value and importance of human factors has not yet been comprehensively accepted throughout the regulated industry. Among the main human factors influencing error in the medical product manufacturing industry, one can include:

- Compliance and quality culture
- Workplace involvement—motivation and attention
- Adequate supervision and staffing
- Procedures and task design
- Training and performance
- Working from memory
- Multitasking

Once we all agree that eliminating all human error is impossible (to do that, we would first have to eliminate all humans), our efforts should address two areas:

1. Reducing the probability of human error from the onset and when an unavoidable error occurs
2. Implementing barriers to detect those human errors and/or to minimize their impact on the quality of our processes

Using the CAPA concepts, these mitigation efforts have two parts:

1. The *preventive* part must encompass important human factors such as better supervision, better procedures and working instructions, and more effective training efforts. Make your processes and your documents as error-proof as you can. Do not hesitate to overuse mistake-proofing features, also known by the Japanese term of *poka-yoke*.
2. For the *reactive* part, you must improve your investigations. Don't accept human error as the root cause; dig into the human factors and think twice before using retraining as a corrective action.

To succeed at error control and reduction, you must consider the influence the following factors have on behavior and performance:

- Design of facilities and equipment

- Information content and format (procedures, work instructions, and job aids)
- Training
- Method of work—supervision and management controls (including adequate resources and clear roles and responsibilities)
- Process of communications

There are three areas in which you must concentrate our effort in order to effectively and dramatically reduce the impact of so-called human error on the bottom line in the regulated industries. To do the right thing the first time, you need:

a. Better, user-centered documents (working instructions, specifications, and procedures) with clear, complete, and comprehensive instructions
b. Better training to ensure that workers understand why they are doing what they are doing, why they must always follow instructions, and what happens when instructions are not followed
c. Better supervision to ensure that workers always follow procedures and working instructions while performing any function in a GMP-regulated environment

Steps in reducing human errors include:

- Addressing the conditions and reducing the stressors that increase the frequency of errors
- Designing facilities and equipment to prevent slips and lapses occurring or to increase the chance of detecting and correcting them
- Driving out complexity and designing jobs to avoid the need for tasks that involve very complex decisions, diagnoses, or calculations, for example, by writing procedures for rare events requiring decisions and actions
- Ensuring proper supervision, particularly for inexperienced staff or for tasks where there is a need for independent checking
- Checking that job aids such as procedures and instructions are clear, concise, available, up-to-date, and accepted by users
- Considering the possibility of human error when undertaking risk assessments

- Thinking about the different causes of human errors during incident investigations in order to introduce measures to reduce the risk of a repeat incident
- Monitoring to ensure that measures taken to reduce error are effective
- Enhancing process (barrier) controls and poka-yokes
- Removing latent failures
- Making people accountable for errors and mistakes within a positive and blame-free environment
- Understanding the hows and whys of human errors

To reduce violations, managers could:

- Take steps to increase the chances of violations being detected using routine monitoring, internal audits, and so on
- Make rules and procedures relevant and practical, and eliminate unnecessary rules or instruction
- Train by explaining the reasons behind certain rules or procedures and their relevance
- Provide more training and better control (for example supervisory presence) for abnormal and emergency situations to minimize exceptional violations

Everyone can make errors no matter how well trained and motivated they are. Sometimes people are "set up" by the system to fail. The challenge is to develop error-tolerant systems and to prevent errors from occurring.

Reducing human error involves far more than taking disciplinary action against an individual. There are a range of measures that provide more effective controls, including design of the job and equipment, procedures, and training. Paying attention to individual attitudes and motivations, design features of the job, and the organization will help to reduce violations. And last but not least important, you must establish an order for error control from best, ideal options to least effective barriers:

- Error or mistake-proof: make the error impossible (computer form cannot be saved if some information is missing; microwave cannot operate if door is open; production tank will unlock and open only if scanned containers of ingredients match the system's bill of material for the lot).

- Error prevention: signals and alarms (for example, the red underline of incorrect spelling in Microsoft Word).
- Minimize the impact of error (inspection and test, or double verification of the addition of components to a mixing tank).

Regulatory Actions: Recalls and Health Hazard Evaluation

Recall activities are one of the clearest examples of the use of risk management activities within our industry. A recall is a method of removing or correcting products that are in violation of laws administered by FDA. Recalls may be conducted on a firm's own initiative, by FDA request, or by FDA order under statutory authority to protect the public health and well-being from products that present a risk of injury or gross deception or are otherwise defective.

A recall is an alternative to an FDA-initiated court action for removing or correcting violative products that have been distributed. 21 CFR 7 sets forth specific recall procedures for FDA to follow in monitoring recalls and assessing the adequacy of a firm's efforts in a recall. A request by FDA that a firm recall a product is reserved for urgent situations and is directed to the firm that has primary responsibility for the manufacture and marketing of the product that is to be recalled.

Recall does not include market withdrawal or a stock recovery. A market withdrawal is a firm's removal or correction of a distributed product that involves a minor violation that would not be subject to legal action by FDA or that involves no violation, for example, normal stock rotation practices, routine equipment adjustments and repairs, and so on. Almost all recalls are conducted on a voluntary basis by the manufacturer.

Health Hazard Evaluation (HHE). An evaluation of the health hazard presented by a product being recalled or considered for recall is conducted by FDA, taking into account, but not necessarily limited to, the following factors:

- Whether any disease or injuries have already occurred from the use of the product.
- Whether any existing conditions could contribute to a clinical situation that could expose humans or animals to a health hazard. Any conclusion shall be supported as completely as possible by scientific documentation and/or statements that the conclusion is the opinion of the individual(s) making the health hazard determination.

- Assessment of hazard to various segments of the population (children, surgical patients, pets, livestock, and so on) who are expected to be exposed to the product being considered, with particular attention paid to the hazard to those individuals who may be at greatest risk.
- Assessment of the degree of seriousness of the health hazard to which the populations at risk would be exposed.
- Assessment of the likelihood of occurrence of the hazard.
- Assessment of the consequences (immediate or long-range) of occurrence of the hazard.

Based on this determination, FDA will assign the recall a classification, that is, Class I, Class II, or Class III, to indicate the relative degree of health hazard of the product being recalled or considered for recall.

Classification. Recalls are classified into a numerical designation (I, II, or III) by FDA to indicate the relative degree of health hazard presented by the product being recalled, as described in Table 7.1.

Recall Strategy. The recalling company must develop a recall strategy that takes into account the following factors as they apply to the individual circumstances of the particular recall:

- Results of health hazard evaluation
- Ease in identifying the product
- Degree to which the product's deficiency is obvious to the consumer or user

Table 7.1 FDA recall classification.

Class I	Class II	Class III
A situation in which there is a reasonable probability that the use of, or exposure to, a violative product will cause serious adverse health consequences or death	A situation in which use of, or exposure to, a violative product may cause temporary or medically reversible adverse health consequences or where the probability of serious adverse health consequences is remote	A situation in which use of, or exposure to, a violative product is not likely to cause adverse health consequences

- Degree to which the product remains unused in the marketplace
- Continued availability of essential products

FDA will review the adequacy of a proposed recall strategy and recommend changes as appropriate. A recalling firm should conduct the recall in accordance with an approved recall strategy but need not delay initiation of a recall pending review of its recall strategy. A recall strategy will address the following elements regarding the conduct of the recall:

1. *Depth of recall.* Depending on the product's degree of hazard and extent of distribution, the recall strategy will specify the level in the distribution chain to which the recall is to extend, as follows:
   - Consumer or user level, which may vary with product, including any intermediate wholesale or retail level
   - Retail level, including any intermediate wholesale level
   - Wholesale level
2. *Public warning.* The purpose of a public warning is to alert the public that a product being recalled presents a serious hazard to health. This is reserved for urgent situations where other means of preventing the use of the recalled product appear inadequate. FDA, in consultation with the recalling firm, will ordinarily issue such publicity. When the recalling firm decides to issue its own public warning, it is requested to submit to FDA its proposed public warning and plan for distribution of the warning for review and comment. The recall strategy will specify whether a public warning is needed and whether it will be issued as:
   - A general public warning through the general news media, either national or local as appropriate
   - A public warning through specialized news media, for example, professional or trade press, or to specific segments of the population such as physicians, hospitals, and so on
3. *Effectiveness checks.* The purpose of effectiveness checks is to verify that all consignees (at the recall depth specified by the strategy) have received notification about the recall and have taken appropriate action. Consignees may be contacted through personal visits, telephone calls, letters, or a combination thereof. A guide entitled *Methods for Conducting Recall Effectiveness Checks* that describes the use of these different methods is available from FDA. The recalling firm will ordinarily be responsible for

conducting effectiveness checks, but FDA will assist in this task where necessary and appropriate. The recall strategy will specify the method(s) to be used and the level of effectiveness checks that will be conducted, as follows:

- *Level A*. 100 percent of the total number of consignees to be contacted
- *Level B*. Some percentage of the total number of consignees to be contacted, which percentage is to be determined on a case-by-case basis, but is greater than 10 percent and less than 100 percent of the total number of consignees
- *Level C*. 10 percent of the total number of consignees to be contacted
- *Level D*. Two percent of the total number of consignees to be contacted
- *Level E*. No effectiveness checks

Firm-Initiated Recall. A firm may decide of its own volition and under any circumstances to remove or correct a distributed product. A firm that does so because it believes the product to be violative is requested to immediately notify the recall coordinator of the appropriate FDA district office. Such removal or correction will be considered a recall only if FDA regards the product as involving a violation that is subject to legal action, for example, seizure. In such cases, the firm will be asked to provide FDA the following information:

1. Identity of the product involved
2. Reason for the removal or correction and the date and circumstances under which the product deficiency or possible deficiency was discovered
3. Evaluation of the risk associated with the deficiency or possible deficiency
4. Total amount of such products produced and/or the time span of the production
5. Total amount of such products estimated to be in distribution channels
6. Distribution information, including the number of direct accounts and, where necessary, the identity of the direct accounts

7. A copy of the firm's recall communication if any has been issued, or a proposed communication if none has been issued
8. Proposed strategy for conducting the recall
9. Name and telephone number of the firm official who should be contacted concerning the recall

FDA will review the information submitted, advise the firm of the assigned recall classification, recommend any appropriate changes in the firm's strategy for the recall, and advise the firm that its recall will be placed in the weekly FDA Enforcement Report. Pending this review, the firm need not delay initiation of its product removal or correction.

A firm may decide to recall a product when informed by FDA that the Agency has determined that the product in question violates the law, but the Agency has not specifically requested a recall. A firm that initiates a removal or correction of its product that the firm believes is a market withdrawal should consult with the appropriate FDA district office if the reason for the removal or correction is not obvious or clearly understood but where it is apparent—for example, because of complaints or adverse reactions related to the product—that the product is deficient in some respect. In such cases, FDA will assist the firm in determining the exact nature of the problem.

Recall Letter. A recalling firm is responsible for promptly notifying each of its affected direct accounts about the recall. The format, content, and extent of a recall communication should be commensurate with the hazard of the product being recalled and the strategy developed for that recall. In general terms, the purpose of a recall communication is to convey:

- That the product in question is subject to a recall
- That further distribution or use of any remaining product should cease immediately
- Where appropriate, that the direct account should in turn notify its customers who received the product about the recall
- Instructions regarding what to do with the product

A recall communication can be accomplished by telegrams, mailgrams, or first-class letters conspicuously marked, preferably in bold red type, on the letter and the envelope "medical device recall [or correction]." The letter and the envelope should also be marked "urgent" for Class I and Class II recalls and, when appropriate, for Class III recalls. Telephone calls or other personal contacts should ordinarily be confirmed by one of the aforementioned methods and/or documented in an appropriate manner. A recall communication should be written in accordance with the following guidelines:

- Be brief and to the point
- Identify clearly the product, size, lot number(s), code(s), or serial number(s), and any other pertinent descriptive information to enable accurate and immediate identification of the product
- Explain concisely the reason for the recall and the hazard involved, if any
- Provide specific instructions on what should be done with respect to the recalled products
- Provide a ready means for the recipient to report to the recalling firm whether it has any of the products (for example, by sending a postage-paid, self-addressed postcard or by allowing the recipient to place a collect call to the recalling firm)

The recall communication should not contain irrelevant qualifications, promotional materials, or any other statement that might detract from the message. If necessary, follow-up communications should be sent to those who fail to respond to the initial recall communication. It is highly recommended that the recalling firm discuss the recall letter with the FDA district office recall coordinator prior to issuing the notification. Consignees who receive a recall communication should immediately carry out the instructions set forth by the recalling firm and, where necessary, extend the recall to its consignees in accordance with the instructions described previously.

Recall Status Reports. The recalling firm is requested to submit periodic recall status reports to the appropriate FDA district office so the agency may assess the progress of the recall. The frequency of such reports will be determined by the relative urgency of the recall and will be specified by FDA in each recall case; generally, the reporting interval will be between two and four weeks. Unless otherwise specified or inappropriate in a given recall case, the recall status report should contain the following information:

1. Number of consignees notified of the recall, and date and method of notification
2. Number of consignees responding to the recall communication, and quantity of products on hand at the time it was received

3. Number of consignees who did not respond (if needed, the identity of nonresponding consignees may be requested by FDA)
4. Number of products returned or corrected by each consignee contacted, and the quantity of products accounted for
5. Number and results of effectiveness checks that were made
6. Estimated time frames for completion of the recall

Recall status reports are to be discontinued when the recall is terminated by FDA.

Termination of a Recall. A recall will be terminated when FDA determines that all reasonable efforts have been made to remove or correct the product in accordance with the recall strategy and when it is reasonable to assume that the product subject to the recall has been removed and proper disposition or correction has been made commensurate with the degree of hazard of the recalled product. Written notification that a recall has been terminated will be issued by the appropriate FDA district office to the recalling firm. A recalling firm may request termination of its recall by submitting a written request to the appropriate FDA district office stating that the recall is effective in accordance with the criteria set forth and by accompanying the request with the most current recall status report and a description of the disposition of the recalled product.

Public Notification of Recall. FDA publishes a weekly FDA Enforcement Report (Food and Drug Administration 2023b) that contains all enforcement actions, including recalls, field corrections, seizures, and injunctions.

Benefit-Risk Determination

FDA has been working extensively for several years to develop what the agency calls "benefit–risk assessment frameworks" covering medical products. Following is a brief discussion of the current state of these initiatives for drugs and biological products and for medical devices.

Drugs and biological products. The U.S. FDA Safety and Innovation Act of 2012 (FDASIA) requires FDA to implement a structured benefit–risk framework in the new drug approval process. By establishing a structured framework for visualizing the comparative benefits and risks and incorporating it into FDA's review templates, FDA and sponsors can understand and communicate more effectively the judgments that went into an FDA product approval decision.

FDA is currently implementing a structured benefit–risk assessment framework to organize its review of a product's benefit and risk, and communicate benefit–risk assessment decisions at the point of approval or post-market regulatory decision. A key element of the 2012 passage of FDASIA was the requirement for FDA to establish a structured benefit–risk framework to improve the transparency and clarity of FDA's benefit–risk decisions and to communicate more effectively the subjective value judgments underlying benefit–risk determinations. FDASIA directs FDA to "implement a structured risk–benefit assessment framework in the new drug approval process to facilitate the balanced consideration of benefits and risks, a consistent and systematic approach to the discussion and regulatory decision-making, and the communication of the benefits and risks of new drugs."

In February 2013, FDA released a draft five-year plan, PDUFA V Implementation Plan: Structured Approach to Benefit–Risk Assessment in Drug Regulatory Decision-Making, which explains that fundamental to FDA's review of new products are a set of clinical judgments and facts regarding the balance of benefits, risks, unmet medical need, social and behavioral science, and scientific uncertainty:

> A framework for benefit–risk decision-making that summarizes the relevant facts, uncertainties, and key areas of judgment, and clearly explains how these factors influence a regulatory decision, can greatly inform and clarify the regulatory discussion. Such a framework can provide transparency regarding the basis of conflicting recommendations made by different parties using the same information. When the final decision is made, a single framework provides a standardized, predictable, and accessible form that communicates the basis for FDA's regulatory decision to the public, while also documenting the decision for reference as FDA considers similar benefit–risk assessments in the future.

The framework is characterized as a 5 × 3 grid that reviewers will fill in with a lay-friendly description of key benefit–risk determinations, including an analysis of the condition, available treatment options, clinical benefits, risks, and potential approaches to minimize risk (see Figure 7.1). The first two domains of the framework, "Analysis of the Condition" and "Current Treatment Options," are informed by incorporating patient views of their own condition and care. Feedback received during FDA's Patient Focused Drug Development meetings is being used to help fill in these domains and is included in the "Voice of the Patient" reports that capture the meeting proceedings.

Decision factor	Evidence and uncertainties	Conclusions and reasons
Analysis of condition		
Current treatment options		
Benefit		
Risk		
Risk management		
Benefit–risk summary assessment		

Figure 7.1 FDA's structured benefit–risk framework.

Medical Devices. Two major guidances were published by U.S. FDA during 2016 regarding benefit–risk determination for medical devices, detailing its approach to considering the benefits and risks for compliance and enforcement actions that could impact medical device availability:

- Guidance for Industry and Food and Drug Administration Staff—*Factors to Consider When Making Benefit–Risk Determinations in Medical Device Premarket Approval and De Novo Classifications,* issued August 24, 2016. (Food and Drug Administration 2016c)
  - The concepts discussed in this guidance are applicable to the medical device development process from design to market. As such, the benefit–risk factors set out herein should be considered during the design, nonclinical testing, pre–Investigational Device Exemption (IDE), and IDE phases, as well as in assembling and assessing PMA applications or de novo requests.
- Draft Guidance for Industry—*Factors to Consider Regarding Benefit–Risk in Medical Device Product Availability, Compliance, and Enforcement Decisions,* issued June 16, 2016.
  - This draft guidance is intended to harmonize FDA's approach to weighing benefits and risks for medical device product availability, compliance, and enforcement decisions with FDA's benefit–risk framework for assessing medical device marketing and investigational device exemption (IDE) applications.
    The benefit–risk factors in this draft guidance also support assessment of medical devices with real world evidence. While the benefit–risk factors in this draft guidance are not identical to the other frameworks, this draft guidance builds upon FDA's

> premarket review benefit–risk policy in an effort to improve consistency in our patient-centered approach and decision making across the total product life cycle.

This draft guidance also mentions that ISO 14971 is an FDA-recognized standard, and assuring conformity with this standard may help device manufacturers meet the requirements specified in the design controls section (21 CFR 820.30) and other sections of 21 CFR Part 820. Both ISO 14971 and 21 CFR Part 820 take a total life-cycle approach to management of risks associated with medical devices, and expect that manufacturers will incorporate post-market data into their device risk management process, including new risks and changes to existing risks identified after a device is on the market.

This draft guidance document provides a benefit–risk framework for FDA and stakeholders regarding use of benefit–risk information in medical device product availability, compliance, and enforcement decisions. Good documentation of risk management decisions by manufacturers may help to streamline these decisions for both FDA and manufacturers, produce outcomes for patients that deliver the most benefit for the least amount of risk, and provide a reasonable assurance of safety and effectiveness.

In the August 24, 2016 guidance, FDA establishes that although a great deal of emphasis is placed on the importance of clinical data in demonstrating the safety and effectiveness of a medical device, nonclinical data also can be critical to understanding a device's safety and effectiveness. Medical devices often have attributes that cannot be tested using clinical methods alone and that play a major role in the safety or effectiveness of the device.

Both clinical and nonclinical data can play a role in FDA's benefit–risk determinations, and the factors discussed in this guidance are informed by both types of data. FDA relies on valid scientific evidence in making risk and benefit determinations, including the critical issue of identifying "probable risks" and "probable benefits" in the first place. In general, probable risks and probable benefits do not include theoretical risks and benefits, but instead ones whose existence and characteristics are supported by valid scientific evidence. Generally, isolated case reports, random experience, reports lacking sufficient details to permit scientific evaluation, and unsubstantiated opinions are not regarded as valid scientific evidence to show safety or effectiveness. However, such information may be considered in identifying a device that has questionable safety and effectiveness, as required in 21 CFR 860.7(c)(2).

In this guidance, FDA details factors considered as part of the benefit–risk determination and also give examples of how the factors interrelate and how they may affect FDA's decisions. Among those factors, the guidance mentions:

1. Assessment of the benefits of devices
    a. Extent of the probable benefit(s)
        i. The type of benefit(s)
        ii. The magnitude of the benefit(s)
        iii. The probability of the patient experiencing one or more benefit(s)
        iv. The duration of effect(s)
2. Assessment of the risks of devices
    a. Extent of the probable risk(s)/harm(s)
        i. Severity, types, number, and rates of harmful events associated with the use of the device
            1. Device-related serious adverse events
            2. Device-related nonserious events
            3. Procedure-related complications
        ii. Probability of a harmful event
        iii. Duration of harmful events
        iv. Risk from false positive or false negative for diagnostics
3. Additional factors in the assessment of the probable benefits and risk of devices
    a. Uncertainty
    b. Patient-centric assessments and patient-reported outcomes
    c. Characterization of the disease
    d. Patient perspectives
        i. Severity of disease or condition
        ii. Disease chronicity
        iii. Availability of alternative treatment/diagnostics option
    e. Availability of alternative treatment or diagnostics
    f. Risk mitigation
    g. Post-market data
    h. Novel technology addressing unmet medical need

CAPA System

Application of risk management techniques to the CAPA process was discussed in Chapter 5. Risk must be used to decide the depth and magnitude of the nonconformities and deviation investigations, and such prioritization needs to be based on the risk and the criticality (magnitude) of the issue. Risk management tools can also provide the basis for identifying, evaluating, and communicating the potential quality impact of a suspected quality defect, complaint, trend, deviation, out-of-specification result, and so on, and to facilitate risk communications and determine appropriate action to address significant product defects in conjunction with regulatory authorities, for example, recall activities.

Following are a few critical points you must remember:

- Always link your investigation and CAPA processes to the risk (importance and magnitude) of the event.
- Provide enough resources to investigate, review, and manage your CAPA system.
- Provide enough time to perform investigation and CAPA activities.
- Professionalize your investigation and CAPA functions (use full-time investigators.
- Establish multifunctional boards for investigation report evaluation and for CAPA plan activities.
- Always consider your investigation and CAPA functions among your most critical processes.

The results of CAPA reviews should reveal any previously unrecognized risks and the effectiveness of risk control measures. This information should also be utilized to determine the effectiveness of the risk management activities and determine required actions to be taken to correct the identified issues and prevent recurrence. For example, a service report indicating a safety-related issue with a medical device is reported to a manufacturer and is determined to be a complaint. The complaint is reviewed and an investigation is initiated. During the investigation, it is determined that a manufacturing process change had occurred. Among the potential causes:

- Inadequate assessment of the process change
- Inadequate (or lack of) revalidation
- Inadequate risk control measures
- Risk control measures not evaluated with the change

For any combination of these, it is expected that the risk management system will use this post-market information to initiate another design risk assessment. The extent of the revised risk assessment will depend on the complaint investigation results. The results of any revised risk assessment should be documented. Any new or revised risk control measures will be part of the overall CAPA activities.

During the investigation of any nonconformity or deviation, a preliminary evaluation of the impact of the event based on the initial data and evidence available is one of the first actions to be performed once a problem is detected. The outcome of that initial impact assessment is a determination of the level of risk. This determination will help us guide the efforts of our investigation process.

At this point, most people lose sight of a tool that has been out there for many decades to help us analyze risk and determine the actions needed to address the risk. That tool is known as failure mode and effects analysis (FMEA). It is a systematic method to identify and prevent product and process problems before they occur. Used proactively, FMEA is an excellent tool of true preventive actions: those actions that can be implemented prior to the first occurrence of the failure.

Most companies do not see the direct link between FMEA and the CAPA system. The main reason for this seems to be the fact that most FMEAs are developed by technical groups, process engineers, and so on. Meanwhile, the CAPA system is mostly led by the quality group. If these teams work as silos, they will never interact in the way they are supposed to interact. A proactive FMEA can feed the CAPA system with true preventive actions, that is, actions that can be implemented prior to the first occurrence of the failure. On the other hand, the CAPA system provides corrective and preventive actions. As can be noticed, both systems provide for a closed-loop process. In many regulated companies, the loop is not closed. For example, in many cases the only interaction between FMEA and CAPA starts with a nonconformance. At that moment, the typical question is "Was that failure mode included in the FMEA?" Unfortunately, the most common answer to this question is "No." The typical action is to revise the FMEA to include that failure mode. However, such action is reactive rather than proactive.

In an effective CAPA system, the answer to the question "Was this failure mode included in the FMEA?" must be "Yes." Then, the next question should be "Based on the FMEA, which kind of risk level does that failure mode represent: low, medium, or high?" This could be used as the basis for the risk prioritization of investigations.

Supplier Selection and Control

Risk management activities should identify hazards and evaluate risks, including those potentially introduced by suppliers early in the product realization process. Risk management roles and responsibilities of the manufacturer and supplier should be defined as part of the purchasing requirements. In addition, prescribed risk control measures derived from the risk management process during product realization should be included in the purchasing requirements as part of the purchasing information.

Established criteria for selection, evaluation, and reevaluation of suppliers of purchased products and services should also be based on the risk associated with identified hazards related to the purchased products and services as determined during the risk management process.

Increasing industry accountability to prevent harm to consumers is critical to an effective medical product safety system. The cornerstone of a prevention-focused approach in the sourcing of ingredients or components and the manufacturing and distribution of medical products is the implementation of a QMS approach addressing the safety, quality, and security responsibilities of all persons who manufacture products, including starting materials, for sale to U.S. consumers. A QMS builds on cGMP requirements, recognizing that supply chain integrity and product safety and quality must be every manufacturer's responsibility.

To reduce risks, manufacturers should build quality into the manufacturing of their products and implement effective preventive measures at their facility, as well as ensure the implementation of such measures at their suppliers' facilities. Manufacturers should be accountable for assessing the hazards introduced by their operations and those of their suppliers and consignees, implementing the QMS, monitoring for problems before they result in harm to consumers, taking swift corrective actions to prevent recurrence of any hazards that are not effectively managed, and assuring that medical products that leave their facilities are safe and effective.

Supplier initial selection and monitoring should take into account factors such as:

- Existing legal requirements
- Overall compliance status and history of the company or facility
- Robustness of a company's quality risk management activities
- Complexity of the site and complexity of the manufacturing process
- Complexity of the product and its therapeutic significance
- Number and significance of quality defects and product recalls

- Results of previous audits/inspections
- Major changes in buildings, equipment, processes, key personnel
- Experience with manufacturing of a product (for example, frequency, volume, and so on)

Importers are uniquely positioned in the supply chain to make sure foreign-made products, ingredients, and components are safe because they are responsible for bringing those items into the United States. In January 2009, FDA and other federal agencies issued a draft guidance for industry on good importer practices (GIPs) to encourage importers to take proactive steps to prevent harm to consumers from the products they import. The guidance sets out recommendations to importers of products, including FDA-regulated products, on practices and procedures to follow to increase the likelihood that the products they import are in compliance with applicable U.S. safety and security requirements. The GIPs are broadly organized under four guiding principles:

1. Establishing a product safety management program
2. Knowing the product and applicable U.S. requirements
3. Verifying product and firm compliance with U.S. requirements throughout the supply chain and product life cycle
4. Taking corrective and preventive action when the imported product or firm is not compliant with U.S. requirements

The goal of GIPs is to encourage importers to institute practices to identify and minimize risks associated with imported products. The draft guidance generally recommends that importers should know the producer of the foreign products they purchase and any other manufacturers with which they do business, that they understand the products they import and the vulnerabilities associated with these products, that they understand the hazards that may arise during the product life cycle, including all stages of production, and that they ensure proper control and monitoring of these hazards.

Outsourcing

Outsourcing activities is a common situation nowadays within the medical product manufacturing industries. ICH Q10 states that:

> A pharmaceutical company can outsource activities at all stages of the product lifecycle. The pharmaceutical quality system, including the management responsibilities described in this

> section, extends to the oversight and review of outsourced activities. Normally under a contract, the contract giver should be responsible for assessing the suitability and competence of the contract acceptor to carry out the work required. Responsibilities for quality-related activities of the contract giver and contract acceptor should be specified in a written agreement.

A manufacturer may outsource processes (for example, sterilization, validation, manufacturing or packaging processes, testing, design, and so on) but must maintain control over these outsourced processes and products. The manufacturer is responsible for incorporating appropriate risk management activities for these processes and products through planning and by ensuring that risk control measures are appropriately applied. Before the approval and implementation of a change to any outsourced process or product, the manufacturer should:

- Review the change
- Assess whether new risks have been discovered
- Determine whether current and/or new individual residual risks and/or the overall risk is acceptable according to the predetermined existing acceptability criteria

If there are any risk control measures applied to outsourced processes or products, the risk control measures and their importance should be documented within the purchasing data or information and clearly communicated to the supplier.

As stated in numerous laws, regulations, and guidances, *the primary manufacturer is responsible:*

- 21 CFR 200.10(b) *"FDA . . . regards extramural facilities as an extension of the manufacturer's own facility."*
- 21 CFR 211.22(a) *"Quality Control Unit (QCU) is ultimately responsible for approving or rejecting drug products manufactured, processed, packed, or held under contract by another company."*
- 21 USC 351(a)(2)(B) states that *"a drug is considered to be adulterated if the drug is not manufactured in conformance with CGMP."*
  - This applies to finished drug products, drug components, and all sites under contract to manufacture or supply drugs and drug components.
  - The manufacturer and private label distributor can both be held liable.

- 21 USC 301 states that *"introducing or causing the introduction of adulterated drugs into interstate commerce is prohibited."*
- ICH Q9(R1) says regarding outsourcing, *"Members of the supply chain are partners."*
- ICH Q10 says regarding outsourcing:
  - Control and review of all outsourcing activities is an element of a manufacturer's pharmaceutical quality system.
  - The manufacturer is ultimately responsible to ensure that processes are in place to assure the control of outsourced activities and quality of purchased materials.
  - A manufacturer should have adequate procedures for auditing and qualifying facilities prior to outsourcing and throughout the process, and for contract management and supervision.

The analysis of recent tragedies involving contaminated medical and food products reveals what went wrong:

- There was a complete lack of traceability for components.
- Manufacturer could not identify the root cause.
- Manufacturer could not determine where problems started.
- Manufacturer could not fix problem and assure it wouldn't happen again.
- Acceptance of product was based largely on a piece of paper purporting quality (certificate of analysis).
- Little or no testing was performed at receiving sites.
- Test methods that were or would have been performed were unable to detect or signal adulteration.

This author recently worked on one case where a major pharmaceutical company outsourced most cleaning and maintenance activities for buildings and equipment at a manufacturing site. The external company brought in a platoon of workers without any previous experience in FDA-regulated industry and without cGMP training. Problems with documentation, including data integrity issues, were the immediate, and logical, outcome of this situation. These dangerous compliance issues can jeopardize the site compliance records with FDA and other foreign regulators. Both companies are now putting huge efforts (and money) into controlling the risk in the form of numerous corrective and remedial actions.

DOCUMENTS AND RECORDS/ CHANGE MANAGEMENT

Risk management techniques can be very useful during the control and management of changes to documents and records. Change control requires three elements:

a. The description of the proposed change

b. Why the change is needed

c. Why the change is acceptable

Risk situations associated with this element range from inadequate justification of the proposed changes (including lack of supporting studies, validations, and so on) to the use of either unapproved or obsolete documents. It is strongly recommended to establish some kind of criticality matrix to evaluate the impact on product quality of changes to facilities, equipment, materials, manufacturing processes, testing methods, and so on. Based on the result of this risk assessment, additional controls (top management approval as an example) can be added to the change approval process.

Risk management techniques must also be used to determine appropriate actions preceding the implementation of a change, for example, additional testing, (re)qualification, (re)validation, or communication with regulators (preapproval notification, and so on). Finally, risk management techniques can be used to identify high-risk equipment and facilities that need to be maintained under strict change control, as well as other equipment and facilities that can be placed under a simpler change management program.

DATA INTEGRITY

Developing a medical product and bringing it to market involves different actors and activities. A fundamental step is linked to the robustness and accuracy of the data submitted by manufacturers to regulatory authorities. These data must be comprehensive, complete, accurate, and true to assure the quality of studies supporting applications for medicines and medical devices to be put on the market. They also must comply with good manufacturing practices (GMPs), good clinical practice (GCP), and good laboratory practices (GLPs).

Recently, there has been a high level of focus on the increasing number of U.S. FDA Warning Letters, WHO Notices of Concern, and EU

Statements of Non-compliance where false or misleading information has been identified during inspections. Failures in data integrity management can arise either from poor systemic control of data management systems due to a lack of knowledge and human error, or from intentionally hidden, falsified, or misleading data.

Between 2015 and 2016, major regulatory bodies such as the European Medicines Agency (EMA), the U.S. FDA, the World Health Organization (WHO), and the Pharmaceutical Inspection Co-operation Scheme (PIC/S) published guidances on the topic of data integrity and data management.

Data integrity enables good decision-making by manufacturers and regulatory authorities. It is a fundamental requirement of the medical products quality system, applying equally to manual (paper) and electronic systems. Promotion of a quality culture, together with implementation of organizational and technical measures that ensure data integrity, is the responsibility of senior management. It requires participation and commitment by staff at all levels within the company, by the company's suppliers, and by its distributors.

Senior management should ensure that data integrity risk is assessed, mitigated, and communicated in accordance with the principles of quality risk management. The effort and resources assigned to data integrity measures should be commensurate with the risk to product quality and balanced with other quality assurance resource demands. Where long-term measures are identified to achieve the desired state of control, interim measures should be implemented to mitigate risk and should be monitored for effectiveness.

On August 11, 2016, the EMA announced the publication of a new GMP data integrity guidance. Data from testing, manufacturing, packaging, distribution, and monitoring of drugs are used by regulators to review the quality, safety, and efficacy of drugs; therefore, ensuring the integrity and completeness of such data is important. The EMA guidance includes a set of questions and answers developed by the agency's GMP/Good Distribution Practice Inspectors Working Group.

The document addresses the assessment of risk to data integrity, risk management strategies, design and control of electronic and paper-based documentation systems, and ensuring data integrity of outside contractors. EMA stated in a press release that the new guidance is aligned with other guidance issued by regulatory authorities participating in the PIC/S. It appears that regulators are taking a closer look at data integrity industrywide.

FDA released its own data integrity draft guidance in April 2016 (further revised in December 2018), and a string of FDA-issued Warning

Letters for data integrity violations have recently been published on the agency's website. Specifically, from January 2015 to May 15, 2016, 21 out of 28 warning letters given to drug manufacturers have involved data integrity issues (Food and Drug Administration 2016a).

The draft guidance, which relies on numerous prior guidances, contains the following recommendations:

- For recording data, manufacturing, or testing steps, numbered, controlled forms to be issued and reconciled by quality assurance.
- Disclosure to FDA of findings of data integrity violations, and "removing at all levels individuals responsible for (data integrity) problems from cGMP positions."
- Both audit trail review and review of electronic testing by quality assurance prior to batch release.
- Control strategies must ensure that all original laboratory records (papers and electronic records) are reviewed by a person to make certain all test results are appropriately reported.
- Immediate, irreversible recording of electronic testing data (including upon completion of each HPLC testing sequence) as opposed to recording only at the end of the day.

The 21 CFR Part 211 regulations that FDA emphasizes in the draft guidance as forming an integral part of the data integrity requirements include:

- § 211.68 (requiring that "backup data are exact and complete," and "secure from alteration, inadvertent erasures, or loss")
- § 212.110(b) (requiring that data be "stored to prevent deterioration or loss")
- §§ 211.100 and 211.160 (requiring that certain activities be "documented at the time of performance" and that laboratory controls be "scientifically sound")
- § 211.180 (requiring that records be retained as "original records," "true copies," or other "accurate reproductions of the original records")
- §§ 211.188, 211.194, and 212.60(g) (requiring "complete information," "complete data derived from all tests," "complete record of all data," and "complete records of all tests performed")

FACILITIES AND EQUIPMENT

Machines and buildings are two essential elements of any medical product manufacturing site along with the people handling the equipment. Life cycles of facilities and equipment are similar, from initial installation to final decommissioning, and risk management techniques are an integral part of those activities. During the design of facilities and equipment, these techniques help to determine appropriate zones in such areas as, for example:

- Flow of material and personnel
- Minimizing contamination, especially cross-contamination
- Pest control measures
- Prevention of mix-ups
- Open versus closed equipment
- Clean rooms versus isolator technologies for aseptic/sterile products
- Dedicated or segregated facilities/equipment

They can also be used to determine appropriate product contact materials for equipment and containers (for example, selection of stainless-steel grade, gaskets, and lubricants), determine appropriate utilities (for example, steam, gases, power source, compressed air, heating, ventilation, air conditioning, water, and so on), and to determine appropriate preventive maintenance for associated equipment (for example, inventory of necessary spare parts). Table 7.2 describes utilization of risk management for control of facilities and equipment.

Over the past several years, the condition of aging manufacturing facilities has become an area of concern both for medical product manufacturers and regulators worldwide. In the case of the United States, inspectional observations against 21 CFR 211 Subpart C "Buildings and Facilities" are becoming more frequent year after year. Citations of "poor maintenance," very often linked to aging facilities, are included in many inspectional reports and Warning Letters.

Table 7.2 Utilization of risk management in facilities and equipment.

Topic	Activity
Qualification	To determine the scope and extent of qualification of facilities, buildings, and production equipment and/or laboratory instruments (including proper calibration methods)
Hygiene aspects	• To protect the product from environmental hazards, including chemical, microbiological, and physical hazards (for example, determining appropriate clothing and gowning) • To protect the personnel and the environment from hazards related to the product being manufactured
Cleaning	• To differentiate efforts and decisions based on the intended use (for example, equipment dedicated to a single product versus multipurpose, batch versus continuous production) • To determine appropriate cleaning validation limits
Calibration and preventive maintenance	• To set appropriate calibration • To set maintenance schedules, including preventive and predictive maintenance

DESIGN AND DEVELOPMENT

Preproduction activities such as design and development of new medical products are a perfect fit for the use of risk management techniques. Two fundamental references that go deep inside design and development risk are ICH Q8(R2) *Guidance for Industry: Q8(R2) Pharmaceutical Development* (International Conference on Harmonization 2010) and GHTF/SG3/N15R8 *Implementation of Risk Management Principles and Activities within a Quality Management System* (Global Harmonization Task Force 2005). Design and development of a medical product is an evolutionary process, embracing many different elements. Integral to activities associated with this process are the identification and control of risks. Risk controls can be influenced by regulatory, technical, and business considerations, as well as considerations of product safety and functionality, with the goal to reduce risk to an acceptable level while maintaining product feasibility and functionality.

Risk management activities should begin as early as possible in the design and development phase, when it is easier to prevent problems rather than correct them later. For each identified hazard, the risk must be estimated. In risk evaluation, the manufacturer decides whether risk reduction is needed. The results from this risk evaluation, such as the need for risk control measures, then become part of the design input. Risk control measures are part of the design output and are evaluated during design verification. This design input/output/verification cycle will iterate and continue throughout the overall design control process until the residual risks have been reduced to an acceptable level and can be maintained at an acceptable level. The overall effectiveness of risk control measures is confirmed during design validation.

However, relying exclusively on design and development processes to control risk is not sufficient. Even the best design and development processes cannot avoid design output error. Once the product is on the market, risk management activities must be linked to quality management processes, for example, production and process controls, CAPA activities, servicing (in the case of medical devices), and customer feedback. Following are some uses of risk management techniques during the design and development stage:

- To assess the critical attributes of raw materials, solvents, active pharmaceutical ingredients (APIs) and starting materials, components and subassemblies for medical devices, excipients, or packaging materials, including criteria for purchasing selection and control
- To establish appropriate specifications, identify critical process parameters, and establish manufacturing controls (for example, using information from development studies regarding the clinical significance of quality attributes and the ability to control them during processing)
- To decrease variability of critical quality attributes of medical products
- To assess the need for additional studies (for example, bioequivalence, stability)

PRODUCTION AND PROCESS CONTROL

Risk management techniques can be used in production areas to:

- Evaluate the frequency and extent of control testing (incoming, in-process, and finished product). Such analysis can be used, for example, to justify reduced testing under conditions of proven control.

- Evaluate and justify the use of PAT in conjunction with parametric and real-time release.
- Establish the suitability of equipment and the frequency of cleaning, maintenance, and calibration, which should be considered with reference to the risks associated with the process.
- Review and update procedures and work instructions to reflect any appropriate risk control measures.
- Determine appropriate production planning (for example, dedicated, campaign, or concurrent production process sequences).

In the case of medical devices, risk management activities should be used in conjunction with regulatory requirements to establish criteria for traceability and define which devices, components, materials, and work environment conditions require traceability. Points to be considered include:

- Origin of components and materials
- Distribution and location of the device after delivery (to the first consignee)
- Intended use of the device (for example, life sustaining, life supporting, or implantable)
- Probability of failure
- Consequences of failure for patients, users, or other persons

Risk assessment of manufacturing processes, using tools such as FMEA, FTA, HACCP, or HAZOP, can help establish or improve process controls by identifying:

- What can go wrong at each step of the manufacturing process
- The impact of a failure mode on the product
- The likelihood of failures
- Controls to detect and prevent a failure or causes

CAPA system information about production—such as the rate of nonconformities, the rate of rework, scrap, and yield—and other sources of quality data should be evaluated and/or compared against the current risk management output to confirm adequacy and completeness of risk controls.

Trend Analysis and Statistical Process Control

Manufacturers need to continually monitor and analyze data to detect any developing trend. Risk management techniques will help during:

- Selection, evaluation, and interpretation of trend results of data within the product quality review
- Interpretation of monitoring data (for example, to support an assessment of the appropriateness of revalidation or changes in sampling methods)
- The continual improvement process throughout the product life cycle

The analysis of data should demonstrate that the decisions and risk control measures determined within the risk management process are appropriate.

Aseptic Process in the Pharmaceutical Industry

Aseptic processes are probably among the most difficult and demanding of medical product manufacturing processes. Because of the nature of aseptic processes, sterile products produced aseptically present a significantly higher risk to the patient than terminally sterilized (for example, autoclaved or irradiated) products. Because of the high level of risk, an effective quality risk management program is necessary to protect the patient. Operator training and behavior, process validation, plant and equipment maintenance and change control management, and manufacturing process documentation are all critical success factors in the aseptic production arena (Parenteral Drug Association 2008; White 2009).

An effective risk management program aids in the careful control of the process, reducing the risk of contamination as well as wasted effort in controlling insignificant risks. While aseptic processing usually involves filling of final drug product, there are other types of aseptic processes, including aseptic assembly of devices or combination products, aseptic crystallization or aseptic precipitation of drug product to produce a sterile bulk-drug substance, and aseptic formulation of final drug product.

Aseptic processes have in common a high level of risk. A contamination event of a sterile product has an enormous regulatory impact and also will impact the reputation of the manufacturer and its financial viability. cGMP regulations for aseptic manufacture of human and veterinary products mandate that an incident involving product sterility failure or media fill contamination is fully investigated, and also that the manufacturer establish an environmental monitoring program that is properly validated to ensure environmental contaminates are detected.

Aseptic processing requires careful control of the aseptic environment, control of personnel practices and procedures, sterilization of equipment and components, extensive environmental monitoring, and many other controls. The number of controls required and the severe consequences of control failure make aseptic processing one of the highest-risk pharmaceutical processes. Quality risk management is an essential tool in ensuring product quality.

Validations and Revalidations

Effective process validation contributes significantly to assuring medical product quality. The basic principle of quality assurance is that a product should be produced that is fit for its intended use. This principle incorporates the understanding that the following conditions exist:

- Quality, safety, and efficacy are designed or built into the product.
- Quality cannot be adequately assured merely by in-process and finished product inspection or testing.

Each step of a manufacturing process is controlled to assure that the finished product meets all quality attributes, including specifications. Risk management techniques such as FTA, FMEA, HAZOP, and HACCP can be used to identify the scope and extent of verification, qualification, and validation activities (for example, analytical methods, processes, sampling, monitoring, equipment, and cleaning methods). It also can help to distinguish between critical and noncritical process steps to facilitate the design of a validation study.

Process validation and the determination of the need for revalidation may be influenced by the results of risk management activities. Results of process validation or revalidation may identify the need for additional risk control measures, for example, confirming or refining specific process parameters and controls when the source of an identified hazard is process variability. When process changes are undertaken, current risk control measures should be reviewed for suitability. This review should also ensure that no new hazards were introduced. When a high-risk process is under validation, there should be proportionately increased sampling, testing, or more-rigorous acceptance criteria to provide greater assurance of process acceptability.

The 2011 FDA process validation guidance (Food and Drug Administration 2011a) specifically incorporates the following risk management principles:

> A successful validation program depends upon information and knowledge from product and process development. This knowledge and understanding is the basis for establishing an approach to control of the manufacturing process that result in products with the desired quality attributes. Manufacturers should:
>
> - Understand the sources of variation
> - Detect the presence and degree of variation
> - Understand the impact of variation on the process and ultimately on product attributes
> - Control the variation in a manner commensurate with the risk it represents to the process and product
> - The degree of control over those attributes or parameters should be commensurate with their risk to the process and process output. In other words, a higher degree of control is appropriate for attributes or parameters that pose a higher risk.

This guidance also establishes that:

- Risk analysis tools can be used to screen potential variables for DOE studies to minimize the total number of experiments conducted while maximizing knowledge gained.
- Qualification of utilities and equipment can be covered under individual plans or as part of an overall project plan. The plan should consider the requirements of use and can incorporate risk management to prioritize certain activities and to identify a level of effort in both the performance and documentation of qualification activities.
- The sampling plan, including sampling points, number of samples, and the frequency of sampling for each unit operation and attribute. The number of samples should be adequate to provide sufficient statistical confidence of quality both within a batch and between batches. The confidence level selected can be based on risk analysis as it relates to the particular attribute under examination. Sampling during this stage should be more extensive than is typical during routine production.

The European Union's *Guide to Good Manufacturing Practice,* Annex 15 (European Commission 2015) also refers to the use of risk management in validation, stating:

> "A quality risk management approach should be applied throughout the lifecycle of a medicinal product. As part of a quality risk management system, decisions on the scope and extent of qualification and validation should be based on a justified and documented risk assessment of the facilities, equipment, utilities and processes. Retrospective validation is no longer considered an acceptable approach."

Risk management is an essential tool for qualification and validation of aseptic/sterile processes. It provides real benefits to the validation process by identifying risks and ensuring that critical risks are controlled. By focusing on managing risks to the patients and users, medical product manufacturers can ensure that the right resources are applied at the right place at the right time.

Laboratory Control

The disposition and release to customers of medical products is based on tests performed by manufacturers. Most of those tests are performed within in-house quality control laboratories, while others are done by subcontractors. Quality control laboratories (physic-chemical and microbiological) represent one of the most critical areas in any medical product manufacturing site. Either false positive or false negative results represent a big risk for the organization and need to be carefully assessed.

In the case of false positive results, for example, a "phantom" failure of sterility caused by inadequate sample handling, an unnecessary investigation will be generated. Because a sterility failure (for a purportedly sterile product) is a critical situation, a high-level investigation will be generated. Most of the time, no root cause is found, and those investigations are often closed as "inconclusive." On the other hand, an erroneous test result can lead to the distribution to the market of a medical product not meeting its predetermined specifications.

Most of these risk considerations should be covered during test method qualification and validation, which will provide an elevated confidence level on the laboratory capabilities and hence on the results obtained. Risk management techniques must be applied to any failure source that can affect the accuracy and/or reliability of a laboratory result. For some types of medical devices, for example, diagnostic products, this is even more critical because an incorrect medical treatment or therapy can be originated from incorrect clinical laboratory test results. Therefore, the following risk management questions should be answered during the test method validation work:

- What can go wrong during the test?
- How bad is it?
- How often can it happen?
- What can be done to mitigate/reduce the risk?

Packaging and Labeling

There are three main areas where medical product manufacturers can benefit from using risk management tools to evaluate packaging and labeling elements:

- *Design of packages and product literature.* To design the secondary package for the protection of primary packaged product (for example, to ensure product authenticity, label legibility)
- *Selection of container closure system.* To determine the critical parameters of the container closure system
- *Label controls.* To design label control procedures based on the potential for mix-ups involving different product labels, including different versions of the same label

A clear example is the regulatory requirements for so-called over-the-counter (OTC) products, where FDA regulation includes the requirement for tamper-evident packaging of OTC drug products. The reason is to improve the security of OTC drug packaging and help assure the safety and effectiveness of OTC drug products. Among the requirements:

> Each manufacturer and packer who packages an OTC drug product for retail sale shall package the product in a tamper-evident package, if this product is accessible to the public while held for sale. A tamper-evident package is one having one or more indicators or barriers to entry which, if breached or missing, can reasonably be expected to provide visible evidence to consumers that tampering has occurred. To reduce the likelihood of successful tampering and to increase the likelihood that consumers will discover if a product has been tampered with, the package is required to be distinctive by design or by the use of one or more indicators or barriers to entry that employ an identifying

characteristic (e.g., a pattern, name, registered trademark, logo, or picture). For purposes of this section, the term "distinctive by design" means the packaging cannot be duplicated with commonly available materials or through commonly available processes. A tamper-evident package may involve an immediate-container and closure system or secondary-container or carton system or any combination of systems intended to provide a visual indication of package integrity. The tamper-evident feature shall be designed to and shall remain intact when handled in a reasonable manner during manufacture, distribution, and retail display.

(1) In addition to the tamper-evident packaging feature described in paragraph (b)(1) of this section, any two-piece, hard gelatin capsule covered by this section must be sealed using an acceptable tamper-evident technology.

(c) Labeling. (1) In order to alert consumers to the specific tamper-evident feature(s) used, each retail package of an OTC drug product covered by this section (except ammonia inhalant in crushable glass ampules, containers of compressed medical oxygen, or aerosol products that depend upon the power of a liquefied or compressed gas to expel the contents from the container) is required to bear a statement that:

(i) Identifies all tamper-evident feature(s) and any capsule sealing technologies used to comply with paragraph (b) of this section;

(ii) Is prominently placed on the package; and

(iii) Is so placed that it will be unaffected if the tamper-evident feature of the package is breached or missing.

(2) If the tamper-evident feature chosen to meet the requirements in paragraph (b) of this section uses an identifying characteristic, that characteristic is required to be referred to in the labeling statement. For example, the labeling statement on a bottle with a shrink band could say "For your protection, this bottle has an imprinted seal around the neck."

Materials

Control of raw materials and components is an essential step to maintaining safety of medical products (Beck et al. 2009). Thorough knowledge of raw materials can mitigate the potential for contamination (microbiological, chemical, and physical). Raw material control for safety also includes identification—being able to verify that you have received the correct material—because a mix-up due to the presence of an incorrectly identified material in a manufacturing process could compromise safety. Control of raw materials is essential to ensure lot-to-lot consistency because variation in them can directly affect the variation of both product and process. An effective raw material control program will also ensure consistent supplies. A single source for a vital raw material can be a significant financial and quality assurance risk. If a supplier goes out of business or experiences quality issues, can that raw material be obtained elsewhere? Has a second source been qualified in case the primary source is no longer available? Does the second source have the capacity to meet your needs?

Some uses of risk management techniques for materials are:

- To determine whether it is appropriate to use material under quarantine (for example, for further internal processing)
- To determine the appropriateness of reprocessing, reworking, or use of returned goods
- To assess the adequacy of arrangements for ensuring the maintenance of appropriate storage and transport conditions (for example, temperature, humidity, container design)
- To determine the stability of the raw materials and components
- To determine whether a raw material is new to the process or a result of a change to an existing raw material (if the result of a change, what studies have been executed to assure comparability?)
- To determine if the use of the raw material in a manufacturing environment presents safety and/or handling risks

SERVICING

Servicing is a regulatory requirement for certain medical device products. It means repair and maintenance activities for such devices. When servicing is a specified requirement, information from risk management activities should be considered. Periodic servicing and maintenance as a means to ensure safe functioning of a device can be a method of risk control.

If a certain risk control measure is necessary for a production process, it may also be necessary to apply the same (or similar) risk control measure to the servicing process. When there is a hazard to service personnel, clear instructions need to be included in service manuals or documentation, and appropriate training provided.

Endnote

1. For in-depth coverage of this topic, see Rodriguez-Perez (2023).

8

Risk Management Is More Than FMEA

Since the first notions and requirements for risk analysis of medical products came around, many regulated companies (if not most of them) started to use the failure mode and effects analysis (FMEA) methodology to accomplish those new tasks.

While this tool can have many good uses, there are also several typical misuses. The most notorious (and widely practiced) is to believe that using *only* FMEA for the entire risk management process is appropriate. FMEA is a good risk analysis tool, but alone it cannot sustain the risk management program. During audits and inspections, many regulated companies often present their FMEAs to inspectors as their whole risk management system. A sound and effective risk management program needs definitively more than good FMEAs.

Comment 83 in FDA's Preamble to QSR 21 CFR 820 states that:

> When conducting a risk analysis, manufacturers are expected to identify possible hazards associated with the design in both normal and fault conditions. The risks associated with the hazards, including those resulting from user error, should then be calculated in both *normal and fault* conditions [emphasis added]. If any risk is judged unacceptable, it should be reduced to acceptable levels by the appropriate means, for example, by redesign or warnings. An important part of risk analysis is ensuring that changes made to eliminate or minimize hazards do not introduce new hazards. Tools for conducting such analyses include Failure Mode Effect Analysis and Fault Tree Analysis, among others.

The standard FMEA process addresses only fault condition hazards and not normal condition hazards. Another shortcoming of this tool is that it addresses only single-fault condition hazards. In addition, the standard

FMEA is not the best tool for documenting risks that are not failure modes (Bills, Mastrangelo, and Wu 2015).

ICH Q9(R1): Quality Risk Management, adopted by FDA in 2006 and revised in 2023, summarizes some of the most common risk management tools. Among the tools mentioned is FMEA.

FDA writes in section I.2:

> FMEA provides for an evaluation of potential failure modes for processes and their likely effect on outcomes and/or product performance. Once failure modes are established, risk reduction can be used to eliminate, contain, reduce, or control the potential failures. FMEA relies on product and process understanding. It methodically breaks down the analysis of complex processes into manageable steps. It is a powerful tool for summarizing the important modes of failure, factors causing these failures, and the likely effects of these failures.

Furthermore, it mentions that FMEA can be used to prioritize risks and monitor the effectiveness of risk control activities.

Here are excerpts from two Warning Letters issued by FDA regarding the incorrect application of the FMEA tool during the past decade:

1. *Lack of integration* of the FMEA process with a performance test. An organization received customer complaints about its tubing, which was assembled incorrectly. As part of the investigation, the organization conducted a performance test, the result of which showed that the incorrect assembly did not affect the product.

But the organization did not update its FMEA, which stated that the incorrect assembly compromised the functionality of the device. Thus, it left itself vulnerable to future occurrences of the same issue.

2. *Lack of definition of the FMEA indexes for severity, occurrence, and detectability.* An organization received customer complaints about device failures, so it rushed to develop a reactive FMEA based solely on the major cause of the complaints: contamination.

When it developed the scales to rank severity, occurrence, and detectability, it did not provide a rationale for each scale. The organization only provided the indexes and the calculated RPN.

In summary, FMEA alone is not enough to guarantee either product safety or quality. While FMEA is a proactive tool within an organization's risk management system (mainly used during the risk analysis stage), it is a dynamic component that must be handled correctly using the following tips:

- Perform an FMEA using a teamwork approach.
- Use a systematic method to complete the FMEA with help from a process map and cause-and-effect matrix.
- Prioritize the order in which the process inputs will be evaluated using the FMEA tool.
- Sort RPNs in descending order.
- Identify all potential root causes.
- Identify and implement CAPAs for each root cause.
- Recalculate RPNs after CAPAs are implemented, not before.

Doing so will ensure that your FMEA is as effective as possible and your products are safe in the hands of the people they're meant to help.

Appendix A
General Examples

Quality risk management requires the manufacturer to establish a list of known and foreseeable hazards associated with the medical products produced. Furthermore, the manufacturer must also consider the foreseeable sequences of events that could lead to hazardous situations and, ultimately, harm. A hazard cannot result in harm until such a sequence of events leads to a hazardous situation. At this stage the risk can be assessed by estimating the severity and probability of occurrence of the harm as well as the overall detectability of the hazardous situation. Figure A.1 represents the relationships between the main elements involved in the determination of risk, where P1 is the probability of a hazardous situation occurring, and P2 is the probability of a hazardous situation leading to harm.

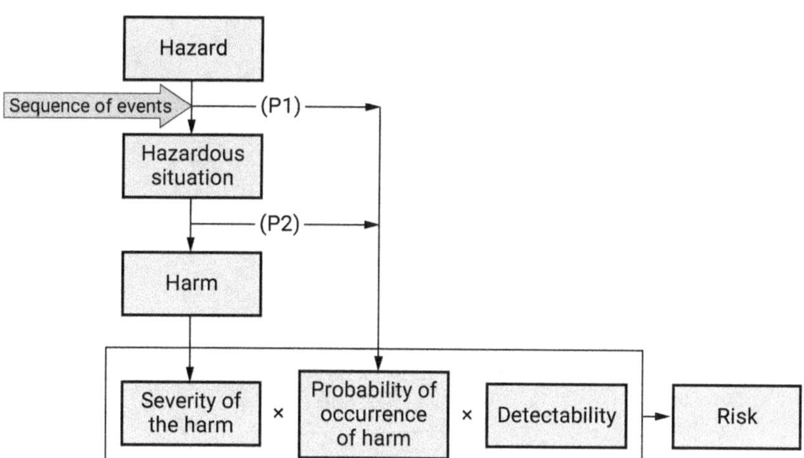

Figure A.1 Elements of risk determination.

The starting point for this assessment should include a review of the risk history of similar types of medical products. Table A.1 is an example of the initial list of hazards. After that, it is possible to initiate the identification of some of the chain of events that along with hazards could result in a hazardous situation and, finally, harm. Analysis and identification can be further complicated by the many initiating events and circumstances that have to be taken into consideration, such as those listed in Table A.2. Table A.3 includes some examples of the relationships between hazards, sequence of events, hazardous situations, and harm (International Organization for Standardization 2019a).

Although this compilation of the list of hazards, hazardous situations, and sequences should be performed as early as possible in the design stage, in practice it is an ongoing activity that will continue throughout post-manufacturing stages. Table A.4 summarizes the attributes and potential applications of the most frequently used risk management tools, while Table A.5 represents an example of qualitative risk component ranking. Finally, Table A.6 contains an example of determination of risk priority ranking based on probability of occurrence and the detectability of the failure mode.

Table A.1 Examples of hazards.

Biological, chemical, and physical hazards	Operational hazards	Information hazards	Energy hazards (medical devices)
Biological • Bacteria • Viruses • Parasites **Biocompatibility** • Allergenicity • Pyrogenicity **Chemical** (exposure to) • Cross-contamination • Cleaning or disinfecting agents • Additives or processing aids • Lubricants **Physical** (exposure to) • Fragment of metal, glass, plastic, or any other foreign materials	**Function** • Incorrect output of the device • Adverse side-effect of the drug • Incorrect data transfer during diagnostic imaging • False positive or false negative result of a clinical diagnostic test **Use error** • Incorrect route of administration for a drug • Incorrect utilization of the medical device	**Labeling** • Incomplete instructions for use • Inadequate description of components • Inadequate description of performance characteristics • Inadequate disclosure of limitations (age, gender, condition, and so on) **Operating instructions** (for medical devices) • Inadequate instructions for pre-use checks (for example, calibration of a blood sugar meter) • Overcomplicated operating instructions • Inadequate instructions for maintenance	**Electromagnetic** • Line voltage • Electric fields • Magnetic fields **Mechanical** • Moving parts • Vibration • Stored energy • Torsion, shear, and tensile force • Moving and positioning of patient **Radiation** • Ionizing • Nonionizing **Thermal** • High temperature • Low temperature

Table A.2 Examples of initiating events.

Category	Initiating events and circumstances
Manufacturing processes	• Inadequate control of changes/document control system • Inadequate control of manufacturing/packaging processes • Inadequate control of subcontractors and suppliers • Inadequate control of materials/components • Inadequate supervision
Storage and shipping methods	• Inadequate packaging • Inappropriate/uncontrolled environmental conditions at storage • Inappropriate/uncontrolled shipping condition
Laboratory controls	• Inadequate test method validation • Inadequate laboratory controls
Environmental factors for medical devices	• Physical (pressure, time, heat, and so on) • Chemical (degradation, contamination, corrosion, and so on) • Electromagnetic fields (susceptibility to electro-magnetic disturbances) • Appropriate power supply • Appropriate lubrication, coolant supply, and so on
Incomplete requirements	Inadequate specifications/requirements for • Design parameters, including formulation and human factors consideration • Operating parameters • Performance requirements • Training of manufacturing/testing personnel • Training of user (if necessary) • Service requirements (for medical devices requiring servicing)
Cleaning, disinfection, and sterilization	• Lack of instructions/specification • Lack of enough details for cleaning, disinfection, and sterilization processes execution • Lack of control of cleaning, disinfection, and sterilization processes

Table A.3 Relationships between hazards, foreseeable sequence of events, hazardous situations, and the harm that can occur.

Hazard	Sequence of events	Hazardous situation	Harm
Biological (microbiological contamination)	1. Inadequate aseptic processing during filling of parenteral drugs 2. Contaminated vial of injectable drug	Bacteria injected into bloodstream of patient	Bacterial infection, sepsis and septicemia, and death
	1. Inadequate instructions provided for decontaminating reused anesthesia tubing 2. Contaminated tubing used during anesthesia	Bacteria released into airway of patient during anesthesia	Bacterial infection, sepsis and septicemia, and death
Chemical	1. Mix-up of drug components 2. Drugs *without* active ingredient	Lack of therapeutic treatment (for example, for an antibiotic or cardiac drug)	Adverse events, including permanent injury or death
	1. Mix-up of drug components 2. Drugs with *incorrect* active ingredient	Unexpected reactions, especially for susceptible patient groups or population	Adverse events, including permanent injury or death (for example, if contaminated with penicillin)
	1. Incomplete cleaning of volatile solvent used in manufacturing of dialysis filters 2. Solvent residue converts to gas at body temperature (during dialysis)	Development of gas bubbles in the bloodstream of patient during dialysis	Gas embolism that could lead to several degrees of brain damage and to death
Electromagnetic energy (line voltage)	Electrode cable plugged into power line receptacle	Line voltage appears on electrodes	Serious burns, heart fibrillation, and death

Continued

Table A.3 *Continued.*

Hazard	Sequence of events	Hazardous situation	Harm
Electro-magnetic energy (ESD: electro-static discharge)	1. Electrostatically charged patient touches infusion pump 2. ESD causes pump and pump alarms to fail 3. Medication not delivered to patient	Failure to deliver medicine unknown to patient. For example, failure to deliver insulin to patient with elevated blood glucose level	Organ damage, decreased consciousness, coma, and even death
Lack of function (no output)	1. Implantable defibrillator battery reaches the end of its useful life 2. Inappropriately long interval between clinical follow-up visits	Device cannot deliver defibrilla-tion shock when an arrhythmia occurs	Death
	1. Inadequate filling process yields some empty capsules of medicine 2. Inadequate QC failed to detect the empty capsules	Patient receives no medication after taking the empty capsule	Adverse events, including permanent injury or death

Table A.4 Risk management tools.

Tool	Description	Application
Basic tools (flowchart, brainstorming, cause-and-effect diagram, check sheet, and so on)	• Simple techniques used to gather and organize data to facilitate decision making	• Process knowledge • Trend analysis
Failure mode and effects analysis (FMEA)	• Assess potential failure modes of processes and their impact on outcomes and process performance • Based on failure modes, risk reduction activities can be developed to control potential failures • Requires strong understanding of the process under evaluation • Outcome is a risk score for each failure mode • Risk score can be further sorted to establish a prioritization scale	• Analyze manufacturing process to identify high-risk steps or critical parameters • Evaluate equipment and facilities to identify critical maintenance activities
Fault tree analysis (FTA)	• Deductive method used to allow analysis of a variety of factors, including human interactions • Very useful for analyzing processes with many interfaces and interactions • Logic analysis of the fault trees helps in identifying simple failure pathways in very complex systems where particular combinations of events that lead to the top event could be overlooked • The "top-down" approach focuses attention on those effects of failure that are directly related to the top event	• Postmortem analysis of deviation, complaints, audit findings, and so on
Hazard analysis and critical control points (HACCP)	• Emphasizes strength of preventive controls • Requires the definition of the critical process parameters • "Bottom-up" approach that evaluates how to prevent hazards from occurring	• Preventive tool rather than reactive/corrective • Assessment of the efficacy of critical control points

Continued

Table A.4 *Continued.*

Tool	Description	Application
Hazard operability analysis (HAZOP)	• Assumption is that events are caused by deviations from design and appropriate operation • Helps to identify potential deviations from normal use of design intentions	• Commonly used to assess process safety hazards
Risk ranking and filtering	• Used to compare and rank risks • Assess multiple qualitative and quantitative factors for each risk by weighing factors and risks	• Tool for prioritization (for example, for audit purposes)

Table A.5 Example of qualitative risk component ranking.

Qualitative ranking	Risk factors		
	Severity	**Occurrence**	**(Lack of) detectability**
High	The impact of the unwanted event is severe	Frequency of occurrence is often	The failure will almost certainly escape detection
Medium	The impact of the unwanted event is moderate	Frequency of occurrence is periodic	Current controls may detect the failure
Low	The impact of the unwanted event is low or negligible	Frequency of occurrence is rare	The failure is very obvious and readily detected

Table A.6 Example of determination of risk priority ranking.

Occurrence		(Lack of) detection: Low	(Lack of) detection: Medium	(Lack of) detection: High
Occurrence	High	The failure is likely to occur, but when it does, it will be detected. It is a *low* or *medium* risk situation.	The failure is likely to occur, and the detection is not certain. It is a *high* risk situation.	The failure is likely to occur, and the detection is not probable. It is a *high* risk situation.
Occurrence	Medium	The failure could occur, but when it does, it will be detected. It is a *low* or *medium* risk situation.	The failure could occur, and the detection is not certain. It is a *medium* or *high* risk situation.	The failure could occur, and the detection is not probable. It is a *high* risk situation.
Occurrence	Low	The failure is not likely to occur, but when it does, it will be detected. It is a *low* risk situation.	The failure is not likely to occur, and if it does, it may be detected. It is a *low* or *medium* risk situation.	The failure is not likely to occur, and if it does,detection is not probable. It is a *medium* risk situation.

Appendix B
Quality Risk Management Case Study

The following case study presents the analysis for the prevention of risks due to potential cross-contamination between drugs, carried out by Business Excellence Consulting Inc. during the period of May 29 to June 2, 2023, at the facilities of XYZ Worldwide Generics Limited located in International Quality Building, Anywhere PR. The material is divided into sections following the logical sequence in which the project was developed.

INTRODUCTION

Whenever production of a single product needs the entire capacity of a facility, issues of potential cross-contamination are of less concern. However, when two or more products are to be manufactured in one facility, one room within a facility, or using common equipment (linked or separate), the potential for cross-contamination becomes a significant issue for consideration. Consistent with the principles embodied in ISPE's Risk-Based Manufacture of Pharmaceutical Products (Risk-MaPP) and FDA's Q9(R1) Quality Risk Management Guidance for Industry, a systematic approach is needed to provide guidance in the identification of risks requiring control strategies to manage them effectively. This should be applied at all stages in the manufacture of pharmaceutical products. Such an approach includes risk assessments that are formally conducted to assist with the identification, implementation, and verification of appropriate risk control strategies as part of the overall risk management process.

During an assessment of XYZ Worldwide Generics Limited's quality risk management procedure, performed by Business Excellence Consulting Inc. on April 18, 2023, it was found that the process to prevent potential cross-contamination between drugs could be further improved. The risk assessment was performed during the period of May 29 to June 2, 2023, by Business Excellence Consulting Inc. It follows a systematic approach, using the FDA's Q9(R1) Quality Risk Management Guidance for Industry, ISPE's Risk-Based Manufacture of Pharmaceutical Products, Six Sigma principles, and quality by design (QbD) concepts.

QUALITY RISK MANAGEMENT MODEL

Quality risk management (QRM) is a systematic process for the assessment, control, communication, and review of risks to the quality of the drug product across the product life cycle. A model for QRM based on FDA's Q9 Quality Risk Management Guidance for Industry is outlined in Figure B.1.

The scope of this case study is to complete the first two steps of the QRM process:

1. Initiate the QRM process
2. Conduct a risk assessment

The other steps are out of the scope of this case study (risk control, risk communication, and risk review) are mentioned only for informational purposes in this section. However, XYZ Worldwide Generics Limited will be responsible for implementing the actions required to complete these steps as part of their QRM process. All the steps in the QRM model, along with the tools used in each one, are described within the next pages.

1. Initiate Quality Risk Management Process

The QRM activities are undertaken by interdisciplinary teams. When teams are formed, they should include experts from the appropriate areas (for example, quality unit, business development, engineering, regulatory affairs, production operations, sales and marketing, legal, statistics, and so on) in addition to individuals who are knowledgeable about the quality risk management process.

The QRM should include systematic processes designed to coordinate, facilitate, and improve science-based decision-making with respect to risk. Some steps used to initiate and plan a QRM process might include the following:

- Define the problem and/or risk question, including pertinent assumptions identifying the potential for risk

- Assemble background information and/or data on the potential hazard, harm, or human health impact relevant to the risk assessment
- Identify a leader and critical resources
- Specify a timeline, deliverables, and appropriate level of decision-making for the risk management process

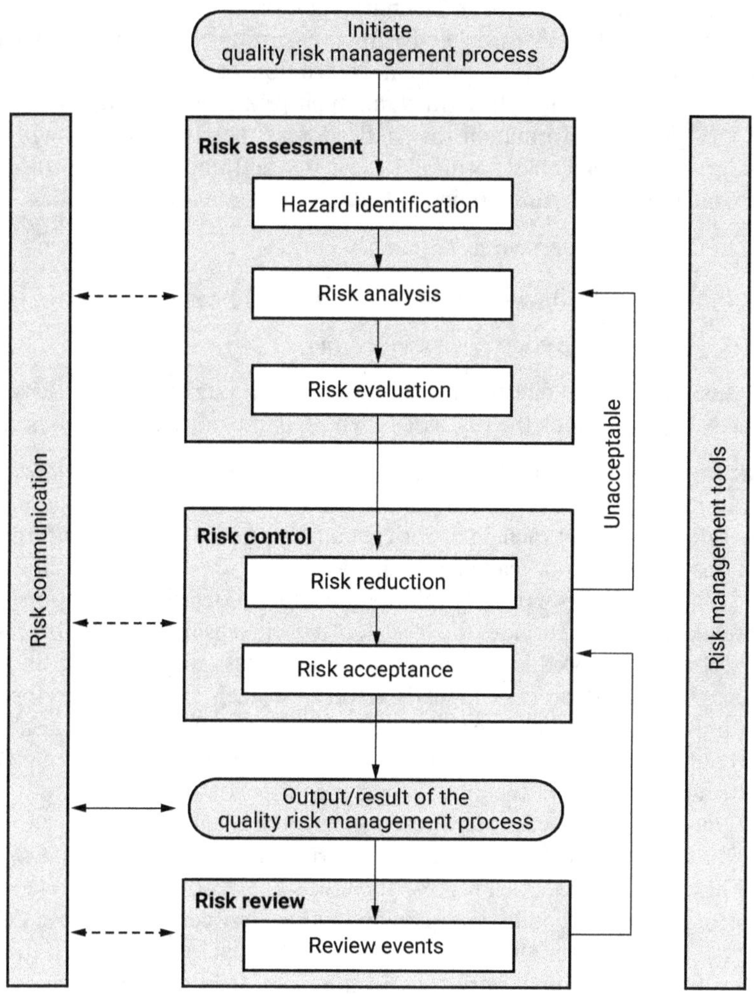

Figure B.1 Overview of a typical quality risk management (QRM) process.

The case study herein starts with a risk management charter, which presents the rationale for this risk assessment. Along with the interdisciplinary team members, the charter includes the business case, problem statement, goal statement, scope, benefits, and project plan. Figure B.2 shows the risk management charter for this case study.

2. Risk Assessment

Risk assessment consists of the identification of hazards and the analysis and evaluation of risks associated with exposure to those hazards. QRM begins with a well-defined problem description or risk question. When the risk in question is well defined, an appropriate risk management tool and the types of information that will address the risk question will be more readily identifiable. As an aid to clearly defining the risk(s) for risk assessment purposes, three fundamental questions are often helpful:

1. What might go wrong?
2. What is the likelihood (probability) it will go wrong?
3. What are the consequences (severity)?

The answers to these questions will be obtained throughout the risk assessment. Now, let us break the risk assessment step into more specific substeps.

Risk Identification. Risk identification is a systematic use of information to identify hazards referring to the risk question or problem description. Information can include historical data, theoretical analysis, informed opinions, and the concerns of stakeholders. Risk identification addresses the "What might go wrong?" question, including identifying the possible consequences. This provides the basis for further steps in the QRM process.

To start the risk identification phase, a process map showing all the process steps and process inputs (factors) for each step was developed. The importance of the process map is to identify all the input variables that could go wrong and have an impact on the product quality. Figure B.3 shows a partial view of the process map for this case study.

Once all the process steps and inputs were identified, each one was analyzed through a cause-and-effect matrix (also called C&E matrix or prioritization matrix). Each process input was analyzed against its relationship to prevent potential cross-contamination between drugs. The C&E matrix is used to filter all the important variables that influence the critical-to-quality (CTQ) output variables; in this case, it is to prevent potential cross-contamination by any of the following routes of exposure: product mix-up, retention in the equipment, mechanical transfer, and airborne

Project Name:	Prevention of cross-contamination between drug products
Date (Last Revision):	2-Jun-23
Prepared By:	Manuel E. Pena, JD, PE
Approved By:	John Doe

Business Case:	Problem Statement:
Whenever production of a single product needs the entire capacity of a facility, issues of cross-contamination are of less concern. However, when two or more products are to be manufactured in one facility, one room within a facility, or by using common equipment, the potential for cross-contamination becomes a significant issue for consideration. Consistent with the principles embodied in ISPE's Risk-Based Manufacture of Pharmaceutical Products and FDA's ICH Q9 guidance, a systematic approach is needed to provide support in the identification of risks requiring control strategies to manage them effectively. This should be applied at all stages in the manufacture of pharmaceutical products. Such an approach includes risk assessments that are formally conducted to assist with the identification, implementation, and verification of appropriate risk control strategies as part of the risk management process.	During an external assessment of the Quality Risk Management process at XYZ Worldwide Generics Limited, held by Business Excellence Consulting Inc. on 18/APR/2023, it was found that the current process to prevent cross-contamination between drug products can be further improved.

Goal Statement:	Project Scope:
Review the risk assessment process to prevent potential cross-contamination between drugs and define the necessary improvements, on or before 30-Jun-23.	Process Start Point: Raw materials and packaging materials warehouse Process End Point: Finished goods warehouse
Expected Benefits: Customers will receive a product free of cross-contamination; XYZ Worldwide Generics Limited will meet the expectations of the guidances about developing a science-based process for risk management.	In Scope: Production areas Out of scope: Office areas

Project Plan:

Task/Phase	Start Date	End Date	Actual End
Plant tour	29-May-23	29-May-23	29-May-23
Develop project charter	30-May-23	30-May-23	30-May-23
Develop process map	30-May-23	30-May-23	30-May-23
Develop C&E matrix	30-May-23	30-May-23	31-May-23
Develop FMEA	31-May-23	1-Jun-23	2-Jun-23
Recommended actions	1-Jun-23	2-Jun-23	2-Jun-23
Risk analysis report	5-Jun-23	5-Jun-23	5-Jun-23

Team:

Name	Role
Manuel E. Pena	Facilitator
John Doe	Manager – QA
Jane Dee	Assistant Manager – QA
Max Fifo	Assistant Manager – Warehouse
Steve Lean	Executive – Production
Mary Shewhart	Deputy Manager – QC
Michael Trane	Assistant Manager – HVAC

Figure B.2 Risk management charter.

Legend:		
	E	Equipment
	S	Step/method
	M	Material
	P	Personnel
	F	Facilities/Environment

Steps	**Raw material and packaging material warehouse**	**Sampling area**	**Dispensing**	**Manufacturing (granulation, mixing, compression, coating, capsule filline, inspection, printing)**	**Primary packing**
Inputs	Decontamination booth (E)	API sampling booth (E)	Barrel pump and hose pipe (E)	Automatic capsule filling (E)	Blister packing machine (E)
	Forklifts (E)	Carts (E)	Cages (E)	Automatic tablet inspection (E)	Bottle airjet cleaner (E)
	Hand pallet trolleys (E)	Excipients sampling booth (E)	Computers (E)	Binder preparation tank (E)	Bottle inverter (E)
	Standard weight trolleys (E)	HEPA filter (E)	Dispensing booth (E)	Canister (E)	Capper (E)
	Weighing balances (E)	Nitrogen purging system (E)	HEPA filter (E)	Capsule deduster/polishing (E)	Change parts (E)
	Documentation (S)	Polybag sealer machine (E)	Nitrogen purging system (E)	Coating solution preparation tank (E)	Check weigher (E)
	API (M)	Pre-filters (E)	Pre-filters (E)	Colloidal mill (E)	Compressed air (F)
	Excipients (M)	Return air risers (E)	Return air risers (E)	Communiting mill (E)	Cotton inserter (E)
	Expired / obsolete material labels (M)	Samples containers (E)	Scoops (E)	Compression machine (E)	Countec machine (E)
		Sampling utensils (E)	Stainless steel containers (E)	Compression machine hopper (E)	Deblistering machine (E)

Figure B.3 Partial view of process map.

transfer. Using this tool, all the process inputs were analyzed. The ratings used for the C&E matrix are illustrated in Figure B.4. A partial view of the unsorted C&E matrix is illustrated in Figure B.5.

This step is completed prior to performing a failure mode and effects analysis (FMEA). By using the C&E matrix, no input is left behind in the risk assessment. The tool considers inputs that have gone wrong and those inputs that might go wrong. Once the C&E matrix is completed, the weighted scores are sorted in descending order and classified as to the relationships between the possible failure of the input variables (factors) and their potential to cause cross-contamination between drugs: strong relationship, moderate relationship, low relationship, or no relationship.

To determine which ranking was assigned, the following rationale was used:

- For weighted scores that have at least one rating of 9, a strong relationship was assigned.
- For weighted scores with no rating of 9 but at least one rating of 3, a moderate relationship was assigned.
- For weighted scores with no ratings of 9 or 3, but at least one rating of 1, a low relationship was assigned.
- For weighted scores with all ratings of zero, no relationship was assigned.

Figure B.6 shows a partial view of the sorted C&E matrix.

Rating	Detectability
0	No relationship with potential cross-contamination
1	Remote relationship with potential cross-contamination
3	Moderate relationship with potential cross-contamination
9	Strong relationship with potential cross-contamination

Figure B.4 Ratings for C&E matrix.

Project Name:	Prevention of cross-contamination between drug products
Date:	30-May-23
Performed by:	Manuel E. Pena, JD, PE
Notes:	

Output Variables (Y's):	Avoid cross contamination by product mix	Avoid cross contamination by retention in the equiment	Avoid cross contamination by mechanical failure	Avoid cross contamination by airborne transfer
Importance Score (1–10):	10	10	10	10

Row Number	Process Step	Process Input	Table of Association Scores (X's to Y's)				Weighted Score
1	Raw material and packaging material warehouse	Decontamination booth (E)	0	3	0	0	30
2	Raw material and packaging material warehouse	Forklifts (E)	0	1	0	0	10
3	Raw material and packaging material warehouse	Hand pallet trolleys (E)	0	1	0	0	10
4	Raw material and packaging material warehouse	Standard weight trolleys (E)	0	1	0	0	10
5	Raw material and packaging material warehouse	Weighing balances (E)	0	1	0	0	10
6	Raw material and packaging material warehouse	Documentation (S)	0	0	0	0	0
7	Raw material and packaging material warehouse	API (M)	9	0	9	0	180
8	Raw material and packaging material warehouse	Excipients (M)	9	0	9	0	180
9	Raw material and packaging material warehouse	Expired / obsolete material labels (M)	0	0	1	0	10
10	Raw material and packaging material warehouse	Packing material (M)	0	0	3	0	30

Figure B.5 Partial view of unsorted C&E matrix.

Project Name:	Prevention of cross-contamination between drug products
Date:	30-May-23
Performed by:	Manuel E. Pena, JD, PE
Notes:	

Output Variables (Y's):	**Avoid cross contamination by product mix**	**Avoid cross contamination by retention in the equiment**	**Avoid cross contamination by mechanical failure**	**Avoid cross contamination by airborne transfer**
Importance Score (1–10):	10	10	10	10

Row Number	**Process Step**	**Process Input**	**Table of Association Scores (X's to Y's)**				**Weighted Score**
7	Raw material and packaging material warehouse	API (M)	9	0	9	0	180
8	Raw material and packaging material warehouse	Excipients (M)	9	0	9	0	180
60	Dispensing	Operators (P)	9	0	9	0	180
144	Manufacturing	Operators (P)	9	0	9	0	180
186	Primary packing	Operators (P)	9	0	9	0	180
191	Primary packing	Staging rooms (F)	9	0	9	0	180
72	Manufacturing	HVAC flows (E)	0	3	0	9	120
120	Manufacturing	Solvent transfer pump (E)	3	9	0	0	120
223	Secondary packing	Operators (P)	9	0	3	0	120
240	Finished goods warehouse	Warehouse personnel (P)	9	0	1	0	100

Figure B.6 Partial view of sorted C&E matrix.

Risk Analysis. Risk analysis is the estimation of the risk associated with the identified hazards. It is the qualitative or quantitative process of linking the likelihood of occurrence and severity of harms. In some risk management tools, the ability to detect the harm (detectability) also factors in the estimation of risk.

From the C&E matrix developed for the risk assessment, those input variables that have a strong relationship with potential cross-contamination between drugs were analyzed through an FMEA. Using this risk assessment tool, each input was analyzed in terms of its potential failure mode. Once the failure mode was identified, the effects of potential cross-contamination between drugs were analyzed in terms of their severity to both the product quality and the patient safety. Then, the potential causes for the failure mode were identified, along with their probability of occurrence. Finally, the current controls (if any) were addressed and evaluated in terms of their effectiveness to prevent or detect the failure mode. The *severity, occurrence,* and *detectability* ratings are illustrated in Figure B.7.

Rating	Severity	Occurrence	Detectability
5	High impact to product quality; Injury to patient	More than once per lot (always)	The current control not capable to prevent or detect the failure mode
4	Moderate impact to product quality; Patient dissatisfaction	Once per lot (frequently)	The action is performed, but no procedure defining how it is performed
3	Low impact to product quality; Likely to result in a complaint	Once at least every three to six months (occasionally)	Procedure exists, is done and registered, but some details are missing
2	Very low impact to product quality; Minor nuisance resulting in no loss	Once every one-to-three years (Rarely)	There is a procedure, the action is performed and recorded
1	No impact to product quality; Be unnoticed and not affect product performance	Once in more than five years (never)	Control is 100% automated

Figure B.7 Ratings for FMEA.

Once all the ratings were assigned, a risk priority number (RPN) was calculated for each of the process inputs using the following formula:

RPN = Severity × Probability of Occurrence × Detectability

A partial view of the unsorted FMEA is illustrated in Figure B.8.

Risk Evaluation. Risk evaluation compares the identified and analyzed risk against given risk criteria. Risk evaluations consider the strength of evidence for all three of the fundamental questions previously mentioned. Once all the RPNs are calculated, they must be sorted in descending order.

The purpose of sorting the RPNs in descending order on the FMEA developed during the risk analysis step is to rank them by their impact on risk. To determine which risk ranking was assigned (high, medium, or low), the following rationale was used: for RPN values lower than 30, a low-risk classification was assigned; for RPN values between 30 and 59, a medium-risk classification was assigned; for those RPN values of 60 or above, a high-risk classification was assigned. Also, regardless of the RPN value, for any potential failure effect with a severity value of 5, a high-risk classification was assigned. A partial view of the sorted FMEA is illustrated in Figure B.9.

The systematic approach used for this risk assessment is based on the article titled "Fail-Safe FMEA" written by José Rodríguez-Pérez and Manuel E. Peña-Rodríguez, published in the January 2012 issue of *Quality Progress* magazine, which is distributed monthly by the American Society for Quality (ASQ).

3. Risk Control

Risk control includes decision-making to reduce and/or accept risks. The purpose of risk control is to reduce the risk to an acceptable level. The amount of effort used for risk control should be proportional to the significance of the risk. Decision-makers might use different processes, including benefit-risk analysis, for understanding the optimal level of risk control.

Risk Reduction. Risk reduction focuses on processes for mitigation or avoidance of quality risk when it exceeds a specified, or acceptable, level. Risk reduction might include actions taken to mitigate the severity and probability of harm. Processes that improve the detectability of hazards and quality risks might also be used as part of a risk control strategy. The implementation of risk reduction measures can introduce new risks into the

Project Name:	Prevention of cross-contamination between drug products
FMEA Team:	See Quality Risk Management Charter
Responsibility:	John Doe
Prepared By:	Manuel E. Pena, JD, PE, Business Excellence Consulting, Inc.

Process									
Process Step	**Process Input**	**Potential Failure Mode**	**Potential Effects of Failure**	**Severity**	**Potential Cause(s) of Failure**	**Probability of Occurrence**	**Current Controls**	**Detection**	**Risk Priority Number (RPN)**
Raw material and packaging material warehouse	API (M)	Wrong labeling	Cross-contamination due to product mix-up	5	Wrong supplies from vendor	4	WH001-26 (Receipt and storage of raw and packing material)	2	40
Raw material and packaging material warehouse	API (M)	Wrong labeling	Cross-contamination due to product mix-up	5	Multiple batches of same material	4	WH001-26 (Receipt and storage of raw and packing material)	2	40
Raw material and packaging material warehouse	API (M)	Wrong labeling	Cross-contamination due to product mix-up	5	Multiple grades of same material	4	WH001-26 (Receipt and storage of raw and packing material)	2	40
Raw material and packaging material warehouse	API (M)	Wrong labeling	Cross-contamination due to product mix-up	5	Wrong pasting of labels	4	WH001-26 (Receipt and storage of raw and packing material)	2	40
Raw material and packaging material warehouse	API (M)	Wrong labeling	Cross-contamination due to product mix-up	5	Illegible labels	4	WH001-26 (Receipt and storage of raw and packing material)	2	40
Raw material and packaging material warehouse	API (M)	Material spillage	Cross-contamination due to mechanical transfer	3	Indequate handling of raw materials	5	WH022-07 (Procedure for handling of spillage of materials); WH040-01 (General Warehouse activities performed by contractual and housekeeping employees)	2	30

Figure B.8 Partial view of unsorted FMEA.

Project Name:	Prevention of cross-contamination between drug products
FMEA Team:	See Quality Risk Management Charter
Responsibility:	John Doe
Prepared By:	Manuel E. Pena, JD, PE, Business Excellence Consulting, Inc.

Process									
Process Step	**Process Input**	**Potential Failure Mode**	**Potential Effects of Failure**	**Severity**	**Potential Cause(s) of Failure**	**Probability of Occurrence**	**Current Controls**	**Detection**	**Risk Priority Number (RPN)**
Manufacturing	Operators (P)	Wrong selection of components	Cross-contamination due to product mix-up	5	Wrong labeling	5	PR022-13 (Manufacturing Procedure)	3	75
Manufacturing	Operators (P)	Wrong selection of components	Cross-contamination due to product mix-up	5	Materials from different batches in single pallet	5	PR022-13 (Manufacturing Procedure)	3	75
Manufacturing	Operators (P)	Wrong selection of components	Cross-contamination due to product mix-up	5	Inadequate training	5	PR022-13 (Manufacturing Procedure); QA013-12 (Training)	3	75
Manufacturing	Operators (P)	Wrong selection of components	Cross-contamination due to product mix-up	5	Work overload	5	PR022-13 (Manufacturing Procedure)	3	75
Manufacturing	Operators (P)	Wrong selection of components	Cross-contamination due to product mix-up	5	Procedure not followed	5	PR022-13 (Manufacturing Procedure); QA013-12 (Training)	3	75
Manufacturing	Operators (P)	Wrong selection of components	Cross-contamination due to product mix-up	5	Wrong documents received	5	PR022-13 (Manufacturing Procedure)	3	75

Figure B.9 Partial view of sorted FMEA.

system or increase the significance of other existing risks. Hence, it might be appropriate to revisit the risk assessment to identify and evaluate any possible change in risk after implementing a risk reduction process.

Risk Acceptance. Risk acceptance is a decision to accept risk. Risk acceptance can be a formal decision to accept the residual risk or it can be a passive decision in which residual risks are not specified. For some types of harms, even the best QRM practices might not entirely eliminate risk. In these circumstances, it might be agreed that an appropriate QRM strategy has been applied and that quality risk is reduced to a specified, or acceptable, level. This specified, or acceptable, level will depend on many parameters and should be decided on a case-by-case basis.

For all those inputs with a high risk described in the sorted FMEA, the company shall identify ways to control the risks (that is, implement risk reduction and risk acceptance activities). For those inputs with moderate risk, the company must decide for which of them risk control actions will be implemented. Then, for those inputs with low risk, no actions might be required at this time because those risks are considered acceptable.

4. Risk Communication

Risk communication is the sharing of information about risk and risk management between the decision-makers and others. Parties can communicate at any stage of the risk management process. The output/result of the QRM process should be appropriately communicated and documented. Communications might include those among interested parties (for example, regulators and industry; industry and the patient; within a company, industry, or regulatory authority). The included information might relate to the existence, nature, form, probability, severity, acceptability, control, treatment, detectability, or other aspects of risks to quality. Communication need not be carried out for each and every risk acceptance. Between the industry and regulatory authorities, communication concerning QRM decisions might be accomplished through existing channels, as specified in regulations and guidances.

5. Risk Review

Risk management should be an ongoing part of the quality management process. A mechanism to review or monitor events should be implemented. The output/results of the risk management process should be reviewed to take into account new knowledge and experience. Once a QRM process has been initiated, that process should continue to be utilized for events that might impact the original QRM decision, whether these events are planned (for example, results of product review, inspections, audits, change control) or unplanned (for example, root cause from failure investigations, recall). The frequency of any review should be based on the level of risk. Risk review might include reconsideration of risk acceptance decisions.

It is recommended to integrate the risk management process with the corrective actions and preventive actions (CAPA) system. Although both systems can operate as separate entities, the best results are obtained when the proactive risk management process is integrated with the CAPA system. In this way, potential failures can be identified and prevented before they do occur. Figure B.10 shows the interrelationship between the risk management process and the CAPA system.

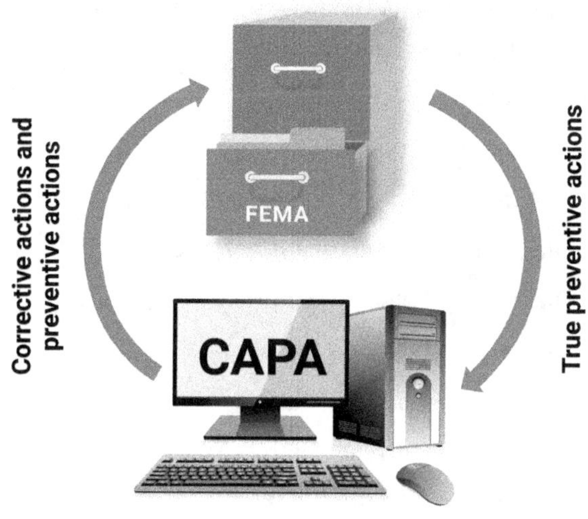

Figure B.10 Relationship between risk management process and CAPA system.

Risk Assessment Results

During the risk assessment stage of the risk management process for potential cross-contamination between drugs carried out at the facilities of XYZ Worldwide Generics Limited, many opportunities for improvement were identified. The risk management process began with the elaboration of a process map, in which the process steps and the input variables (factors) of each of these process steps were identified. Then, through a cause-and-effect matrix, we identified which of the input variables of the different process steps could have a certain type of relationship in terms of potential cross-contamination between drugs. Once those input variables that have a strong and direct relationship with the potential cross-contamination between drugs were identified, they were evaluated in terms of their risk level. For this, an FMEA was developed. All the inputs in the FMEA were evaluated for the risk of potential cross-contamination between drugs. The purpose of the FMEA was to identify the level of risk contributed by each of the analyzed variables, as well as to identify the potential causes of these failures and the current controls to prevent or detect such causes or their potential failure modes.

Using FEMA (including only high- and moderate-risk variables), several Pareto diagrams were developed to analyze the data, find opportunities for improvement, and prioritize the order in which the improvement plans should be established. The Pareto diagrams developed and explained within this case study are:

- Pareto Diagram by Process Step
- Pareto Diagram by Process Input
- Pareto Diagram by Process Input Category
- Pareto Diagram by Potential Failure Effect
- Pareto Diagram by Potential Failure Mode
- Pareto Diagram by Potential Causes
- Pareto Diagram by Current Controls
- Pareto Diagram by Current Controls Type
- Pareto Diagram by Severity Index
- Pareto Diagram by Probability of Occurrence Index
- Pareto Diagram by Detectability Index

For each of the Pareto diagrams included within this case study, a brief narrative is presented about the evaluation of the results obtained by analyzing those diagrams. Likewise, a series of recommendations are presented to mitigate or eliminate the risk of potential cross-contamination between drugs for each of those Pareto diagrams. It is important to emphasize that they are simply recommendations and that XYZ Worldwide Generics Limited will be responsible for identifying the required actions and developing the appropriate plans for its implementation and for the subsequent effectiveness verification of such actions.

ANALYSIS BY PROCESS STEP

Figure B.11 shows the Pareto diagram of those stages of the process for which a high or moderate risk was determined, because they obtained a risk priority number (RPN) equal to or greater than 30 in the FMEA. When analyzing the Pareto diagram, it can be observed that a cumulative 75.1 percent of the total risk is reflected in the areas of:

1. Manufacturing
2. Primary packing
3. Dispensing

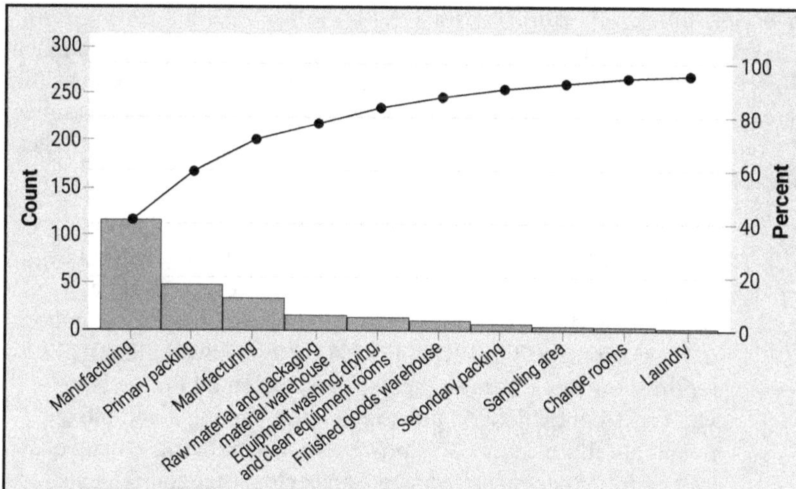

Count	118	45	35	18	15	12	8	6	5	3
Percent	43.9	18.2	13.0	6.7	5.6	4.5	3.0	2.2	1.9	1.1
Cum %	43.9	62.1	75.1	81.8	87.4	91.8	94.8	97.0	98.9	100.0

Figure B.11 Pareto diagram by process step.

This is not uncommon, because at these stages of the process the raw materials are exposed to the work environment; thus, potential cross-contamination can occur through direct contact between the materials. The remaining 24.9 percent corresponds to areas where potential cross-contamination between drugs can occur through indirect contact.

Therefore, the more rigid controls should be established in these areas where potential cross-contamination between drugs can occur through direct contact with the product: manufacturing, primary packing, and dispensing.

ANALYSIS BY PROCESS INPUT

Figure B.12 shows the Pareto diagram categorized by the different input variables (factors) that were identified for all the processes. It can be seen that the most frequent process input is operators (P). Also, the first eight variables include each of the process input types: **M**aterial, **P**ersonnel, **F**acilities/Environment, and **E**quipment.

A more specific Pareto diagram by process input category is presented in Figure B.13. In this figure, the individual process inputs were combined in categories: material, personnel, facilities/environment, and equipment. In this way, those process input categories with the highest risk can be identified. Each of the process input category variables will be explained in more detail.

From Figure B.13, those process inputs (factors) related to equipment (E) represent 52.4 percent of the total, while personnel (P) represents 28.3 percent, material (M) represents 11.9 percent, and facilities/environment (F) represents 5.9 percent of the total. Let us now evaluate each of these categories individually:

1. **Equipment (E):** Most of the potential cross-contamination related to equipment deals with inadequate cleaning caused by procedures not followed or procedures not clear. This type of situation produces potential cross-contamination by the retention in the equipment. The sorted FMEA tables shown in the final report show that for those inputs whose potential failure effect was cross-contamination due to retention in equipment, the potential failure causes are "procedure not followed," "inadequate procedure," "person not trained," and "cleaning requirements not identified." Because XYZ Worldwide Generics Limited has more than 500 different manufacturing equipment and produces over 1,000 different products (tablets and capsules), equipment cleaning is of the utmost importance. It is recommended to review the procedures related to equipment cleaning to ensure they provide enough details. It is also recommended that those cleaning procedures be as visual as possible. In addition, training on procedures (if revised) must be provided to all personnel involved in equipment cleaning operations. Finally, an effectiveness evaluation of those trainings must be included as part of the training process.

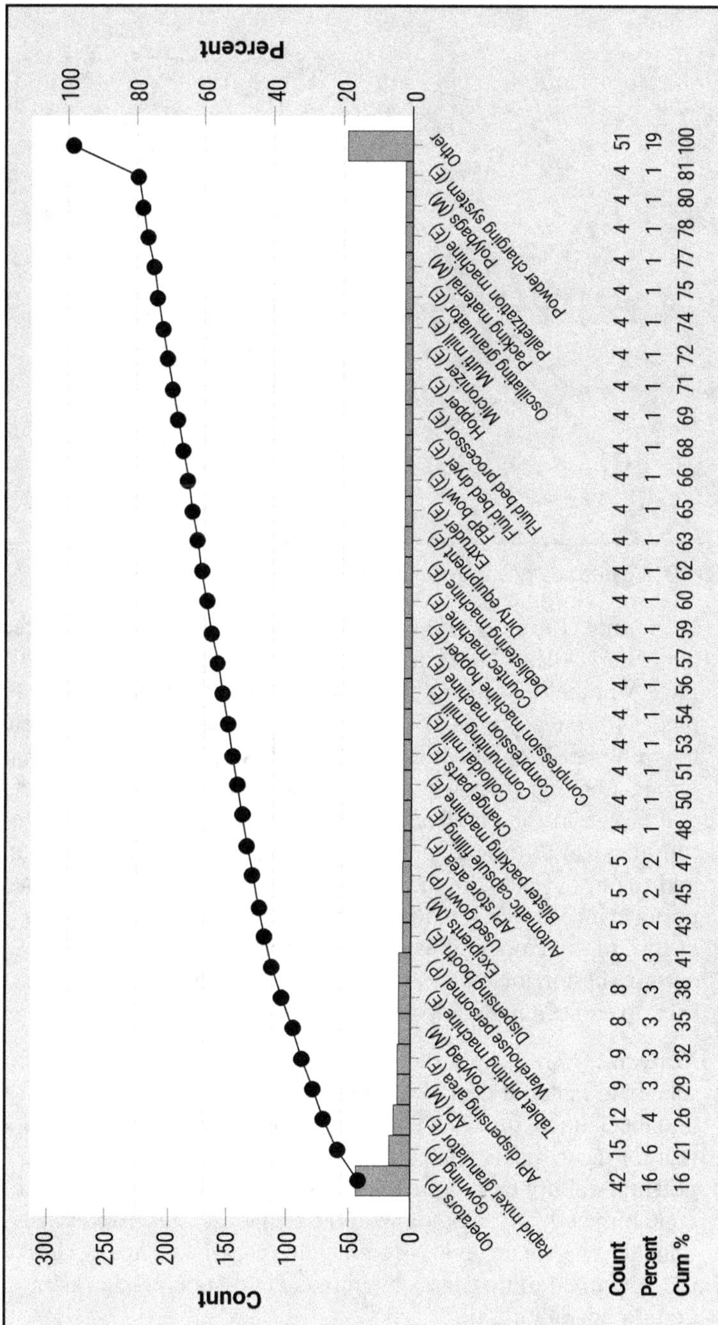

Figure B.12 Pareto diagram by process input.

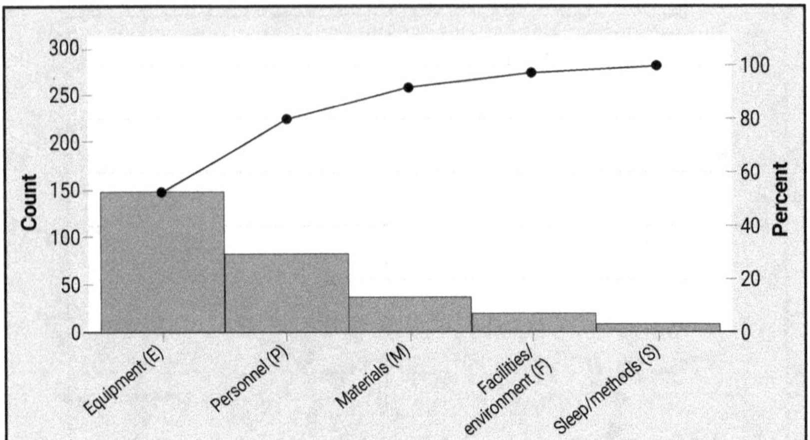

	Equipment (E)	Personnel (P)	Materials (M)	Facilities/ environment (F)	Sleep/methods (S)
Count	141	76	3.2	16	4
Percent	52.4	28.3	11.9	5.9	1.5
Cum %	52.4	80.7	92.6	98.5	100.0

Figure B.13 Pareto diagram by process input category.

2. **Personnel (P):** Most of the potential cross-contamination related to personnel deals with contamination due to product mix-up. The FMEA in the final report shows that most of the potential failure modes for the potential cross-contamination due to mix-up relate to wrong selection of components, dispensing wrong components, wrong labeling, and so on. In most cases, the potential causes of that failure mode are related to human performance: inadequate training, work overload, procedure not followed, lack of alertness, and so on. To deal with those potential causes related to human performance issues, it is recommended to implement double-checks in those operations prone to human errors. Also, make certain these critical tasks where cross-contamination due to mix-up can occur are adequately supervised.

3. **Material (M):** Most of the potential cross-contamination related to material deals with contamination due to mix-up and mechanical transfer. The FMEA in the final report shows that for those inputs related to materials, the most common potential failure modes are related to wrong labeling and use of contaminated polybags. As with the previously discussed issues related to personnel, it is recommended that the company address the potential failure causes by implementing double-checks in certain operations.

4. **Facilities/Environment (F):** Looking at the final report, it can be noted that most potential cross-contamination related to facilities/environment results in mechanical transfer issues. The most common potential failure modes identified with this situation are material spillage and residues from previous product. Most of the potential causes are human-related: inadequate handling of materials, improper cleaning, procedure not followed, and so on. As with the previous process input types, this specific potential failure cause will be addressed with providing adequate supervision and improved training, followed by a training effectiveness evaluation.

ANALYSIS BY POTENTIAL FAILURE EFFECT

Figure B.14 shows the Pareto diagram by potential failure effects. As can be seen in the diagram, 46.8 percent of the high and moderate risk of potential cross-contamination between drugs is associated with potential contamination by retention in the equipment, 28.3 percent is due to mechanical transfer, and the remaining 24.9 percent is related to potential contamination by product mix-up. As mentioned earlier, with more than 500 pieces of manufacturing equipment in the production areas, cleaning becomes an important consideration to avoid potential cross-contamination due to retention in the equipment. Following is a description of each of the potential failure effect categories.

Cross-contamination by *retention in the equipment* is defined as the transfer of those residues of material on the contact surfaces of one product to another, in the same equipment used in a sequential or campaign way. The factors that cause cross-contamination due to retention in the equipment are mainly related to the effectiveness of the cleaning procedures and their application by the related personnel after completing a campaign, and include the cleaning of by-products or residues of cleaning material, as well as waste of the product. The development of adequate cleaning systems is an important area in pharmaceutical manufacturing. To avoid this type of cross-contamination, cleaning procedures must be developed to provide the right cleaning agent under the right conditions for effective cleaning. This requires knowledge of the "cleanliness" of the compound. Studies should be conducted to evaluate the properties required to remove a compound using various surface-cleaning agents and surface finishes. Cleaning processes that have been optimized by selecting the best cleaning agents and cleaning parameters offer the greatest ability to uniformly reduce product waste to

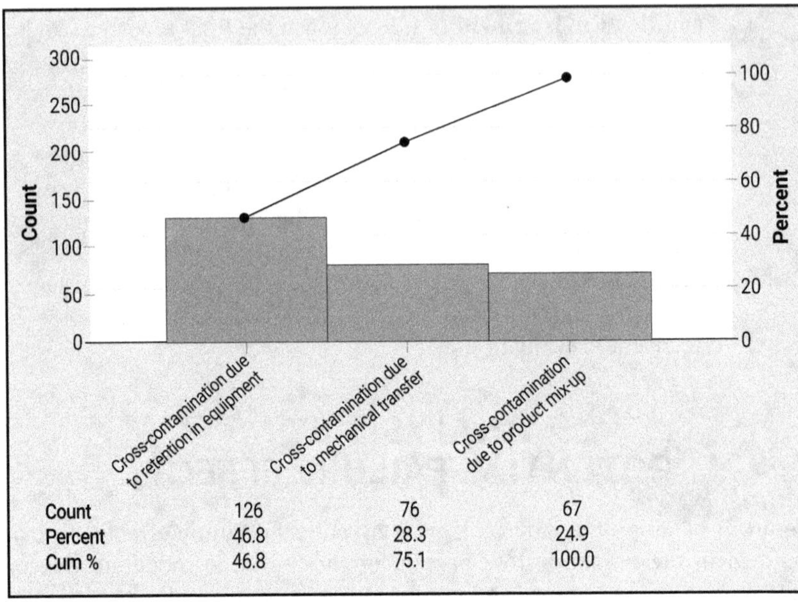

Figure B.14 Pareto diagram by potential failure effect.

the lowest risk levels. Before use, cleaning procedures should be subject to risk assessments to identify and assess the risk of failure, improve them as necessary, and make them more reliable and robust.

On the other side, the cross-contamination by *mechanical transfer* includes contact surfaces of the product that have been contaminated by contact with other contaminated surfaces (for example, in cleaning areas), inadvertent or transient contact with other nondesignated contaminated contact areas of the product, and direct contact of the product with other surfaces, such as staff gowning and/or gloves. The risk of cross-contamination by mechanical transfer can occur with the co-location of a process with contaminated equipment or articles; for example, operators who wear contaminated gowning or contact between process equipment contaminated with different materials. Some factors that usually contribute to cross-contamination by mechanical transfer include:

- Use of common areas where the segregation of processes and equipment is deficient; for example, common areas of manual cleaning where dirty and clean equipment of different processes can be present at the same time
- Use of common gowning rooms

- Use of common equipment; for example, hand tools or exchanging pieces between different processes

The use of certain equipment may cause operators or mobile equipment to become contaminated. This could then potentially transit to other process areas without undergoing decontamination procedures. Issues to be considered include:

- The presence of highly manual, uncontained/open processes
- The specification and location of decontamination procedures for users and equipment
- Routing of equipment and users around facilities especially where there is common routing of product

Finally, cross-contamination by *product mix-up* is frequently cited by regulators as one of the most common causes for regulatory action and/or product recall. Product mix-up refers to the contamination at unsafe levels of one product with another. The highest risks from mix-up normally stem from failures due to basic human failures (for example, not following procedure) or system weakness (for example, poor material labeling). Obviously, there is an increased potential risk of cross-contamination if two or more products are being handled in the same area, even with appropriate batch to batch segregation. With a catalog of more than 1,000 different products at XYZ Worldwide Generics Limited, the potential of cross-contamination due to mix-up takes on important relevance. The risk analysis to identify potential cross-contamination routes should include a review of basic practices to confirm that appropriate systems provide a high level of assurance that mix-up will not occur, such as:

- Identification and status labeling and controls of materials and products
- Status labeling and controls of equipment and facilities
- Receipt and controls of materials and products
- Physical segregation, process flow, and security during material and product handling, storage, and staging

The actions necessary to avoid each of these routes of potential cross-contamination will be addressed specifically during the risk control phase of the QRM process by XYZ Worldwide Generics Limited. An action plan, including the actions, responsible person, target completion dates, and actions taken, will be part of the FMEA file. Once those actions are implemented and their effectiveness is evaluated, the severity, occurrence, and detectability indices must be re-evaluated and the RPNs recalculated. This process must be continuously revised, especially when new process changes are implemented and whenever new risks are identified.

ANALYSIS BY POTENTIAL FAILURE MODE

Figure B.15 shows the Pareto diagram by potential failure modes. As can be seen in the diagram, the most frequent potential failure mode is "Equipment not properly cleaned" (46 percent), followed by "Contaminated gowning" (10 percent) and "Wrong selection of components" (9 percent).

These three potential failure modes are directly linked to the three potential failure effects presented in Figure B.14 (potential cross-contamination by retention in the equipment, by mechanical transfer, and by product mix-up).

When all the potential failure modes presented in Figure B.15 are analyzed in general terms, most of them are directly related to human factors, either due to not following the corresponding procedures (SOPs)

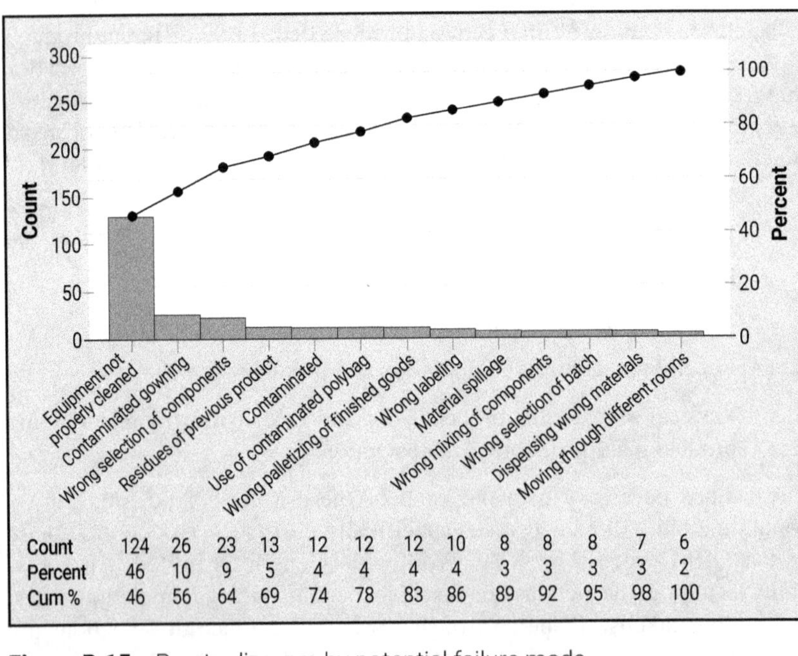

Count	124	26	23	13	12	12	12	10	8	8	8	7	6
Percent	46	10	9	5	4	4	4	4	3	3	3	3	2
Cum %	46	56	64	69	74	78	83	86	89	92	95	98	100

Figure B.15 Pareto diagram by potential failure mode.

or inadequate cleaning of the equipment. Therefore, we recommend that XYZ Worldwide Generics Limited reinforce training for all staff about the importance of proper execution of established procedures, as well as on the adequate cleaning of work areas and equipment.

ANALYSIS BY POTENTIAL CAUSES

Figure B.16 shows the Pareto diagram for potential causes. The first four potential failure causes represent a cumulative 56 percent of the total causes.

These causes are listed below, in the descending order in which they appear in the Pareto diagram:

1. Procedure not followed (20 percent)
2. Inadequate procedure (13 percent)
3. Person not trained (12 percent)
4. Cleaning requirements not identified (12 percent)

Out of the top four potential causes, three of them deal with procedures (procedure not followed, inadequate procedure, and cleaning requirements not identified). Therefore, it is recommended that XYZ Worldwide Generics Limited evaluate cleaning procedures and make sure to establish the appropriate controls to avoid these potential failure causes. Among those controls, operator training is imperative because the potential failure cause called "Person not trained," which is the third potential cause in descending order in the Pareto diagram, accounts for 12 percent of the total potential causes. Once the controls are established, the company is required to measure the effectiveness of these controls. Finally, it is recommended that XYZ Worldwide Generics Limited update the table of the FMEA and reassign the occurrence and detectability indexes to recalculate the RPNs once the actions have been implemented and their effectiveness has been verified.

From the sorted FMEA table in the final report, it can be seen that for those factors with RPN values above 60, the current controls index assigned to all of them was 3, which was defined as "Procedure exists, is done and registered, but details are missing." In other words, procedure exists and is executed, but the procedure needs some improvement. By updating the procedures and providing adequate training and supervision, the current controls index could be lowered to a value of 2, which means "There is a procedure, the action is performed and recorded." Therefore, the new RPN for those factors would be 50, which means that risk has been reduced from high to moderate.

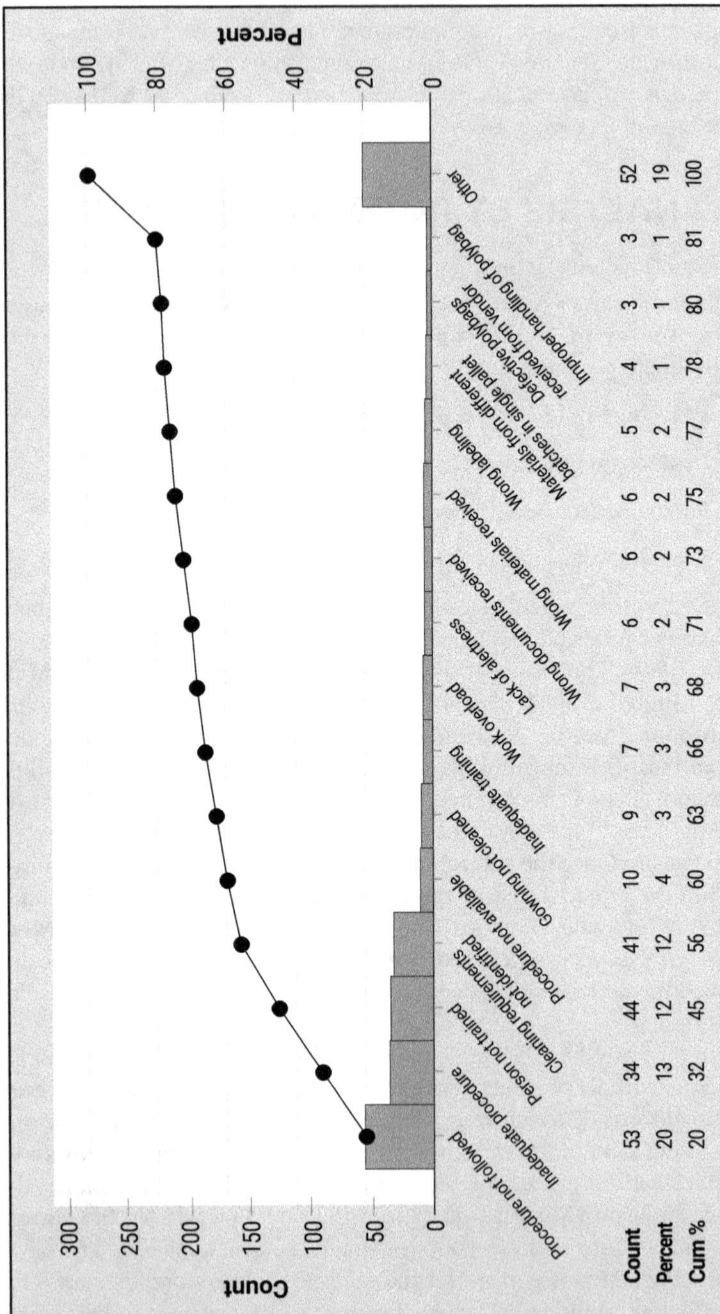

Figure B.16 Pareto diagram by potential causes.

ANALYSIS BY CURRENT CONTROLS

Figure B.17 shows the Pareto diagram for current controls. It can be noted that the three most common procedures mentioned through the FMEA are the following:

- QA013-12 – Training
- PR152-18 – Packaging procedure
- PR005-13 – Entry and exit from production block and process area for factory personnel

Precisely, these controls are directly linked with some of the most common potential causes identified in Figure B.16. In fact, an adequate implementation of QA013-12 will be fundamental in reducing the potential failure cause related to "Person not trained."

When looking at Figure B.17, many procedures are listed one, two, or three times. This makes it difficult to determine if any specific type of procedure is referenced more than others. To overcome that confusion, Figure B.18 shows the same current controls but presented by procedure type. The procedures were classified in one of the following categories:

- Production (PR)
- Quality assurance (QA)
- Warehouse (WH)
- Finished goods (FG)
- Human resources (HR)

From Figure B.18, 71 percent of the procedures listed in the FMEA refer to production, most of them related to the cleaning of a specific equipment. The 16.5 percent related to quality assurance is just one procedure: QA013-12, Training procedure. Another 8.1 percent of the procedures are related to warehouse. These procedures are related to: receipt and storage of raw and packing material, dispensing of raw materials, entry and exit procedure to sampling area, and handling of spillage of materials. It is obvious that production procedures are, by far, those with the highest impact on preventing potential cross-contamination between drugs. It is recommended to revise all procedures related to cleaning, especially those related to equipment cleaning. It is expected that after all procedures are revised, training on these procedures given to all involved, and effectiveness evaluation of the training performed, the risks related to inadequate equipment cleaning can be reduced to accepted levels.

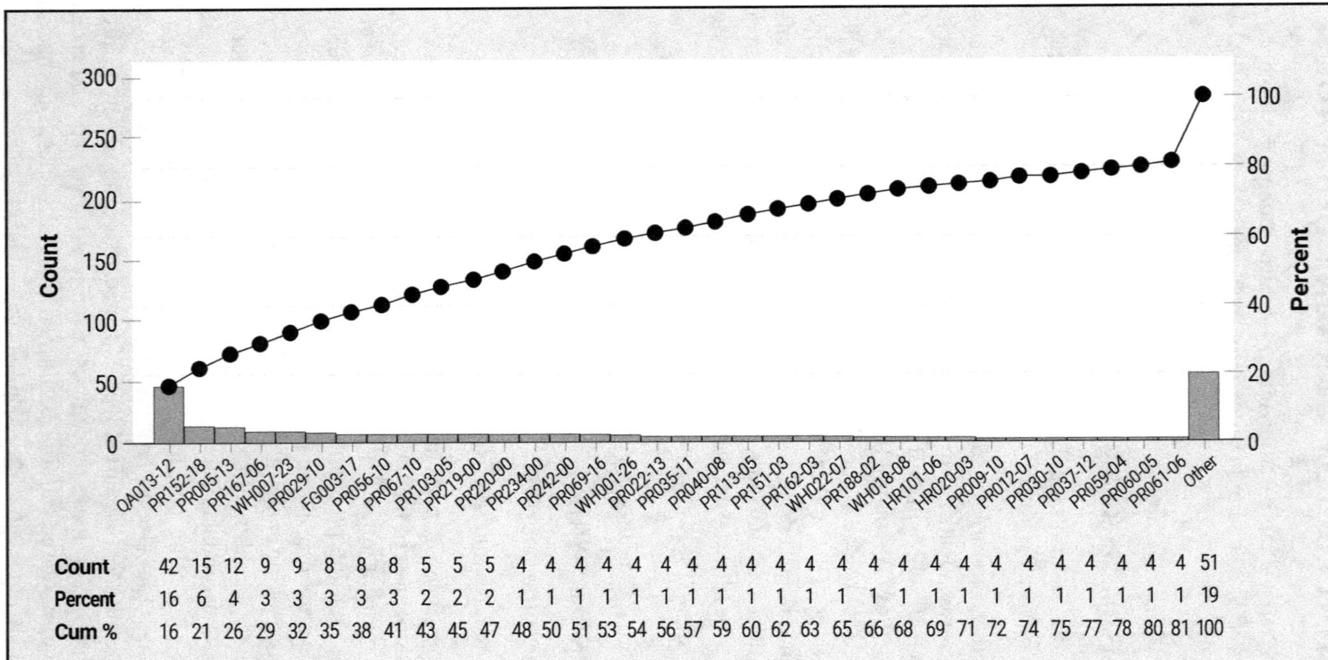

Figure B.17 Pareto diagram by current controls.

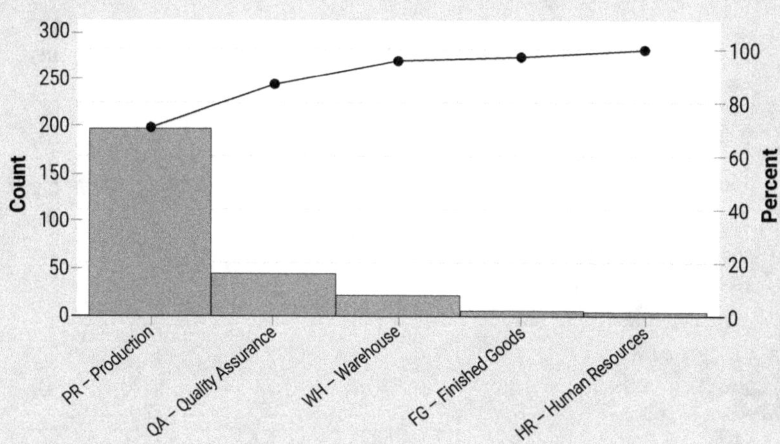

Count	352	82	40	12	10
Percent	71.0	16.5	8.1	2.4	2.0
Cum %	71.0	87.5	95.6	98.0	100.0

Figure B.18 Pareto diagram by current control type.

The Pareto diagrams presented in the previous sections were developed using the information related to those situations for which a high or moderate risk was determined, because they obtained an RPN equal to or greater than 30. On the other side, the Pareto diagrams presented next used the data for all types of risk: high, moderate, and low. In these diagrams, the severity, occurrence, and detectability indices assigned during the development of the FMEA will be analyzed.

ANALYSIS BY SEVERITY INDEX

Figure B.19 shows the Pareto diagram by severity index. The following scale was used to classify the severity as part of the FMEA:

5 = High:	High impact to product quality; Injury to patient
4 = Moderate:	Moderate impact to product quality; Patient dissatisfaction
3 = Low:	Low impact to product quality; Likely to result in a complaint
2 = Very low:	Very low impact to product quality; Minor nuisance resulting in no loss
1 = No severity:	No impact to product quality; Be unnoticed and not affect product performance

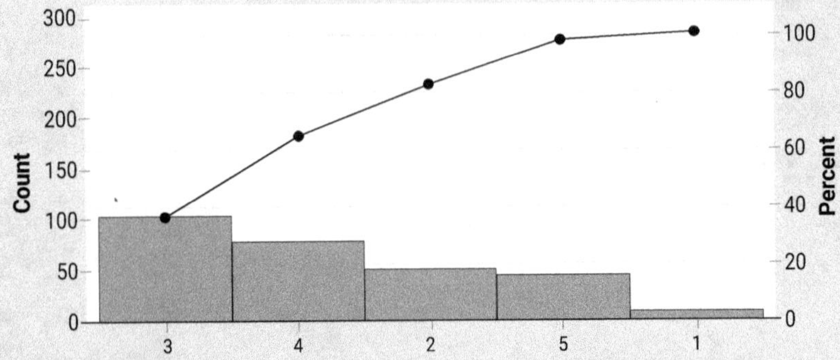

	3	4	2	5	1
Count	178	135	88	77	12
Percent	36.3	27.6	18.0	15.7	2.4
Cum %	36.3	63.9	81.8	97.6	100.0

Figure B.19 Pareto diagram by severity index.

The Pareto diagram shows that in a cumulative 81.8 percent of the potential failure effects, the severity is moderate (4), low (3), and very low (2). A total of 15.7 percent of potential failure effects have a high severity (5). Finally, in 2.4 percent a classification of no severity (1) was assigned. All the potential failure effects that were ranked with a severity index of 5 were related to cross-contamination due to product mix-up. This is the riskiest type of cross-contamination in this facility and could be caused by either mixing the wrong components in the production area or the wrong selection of components in the secondary packaging area. As previously mentioned, with a catalog of more than 1,000 different products manufactured at XYZ Worldwide Generics Limited, this type of potential cross-contamination takes on important relevance.

In most cases where a potential cross-contamination due to retention in the equipment and mechanical transfer could occur, a severity index of 3 or 4 was assigned. This was because these are not penicillin-related products, in which the severity of this type of cross-contamination could produce more direct and adverse effects on the patient. Finally, for those cases where the potential cross-contamination could be related to airborne transfer, a severity index of 2 was assigned. The rationale is because in case such cross-contamination occurs, only minor residues of products could contaminate other products. Because these are not penicillin-related products, the severity of such cross-contamination would be lower.

ANALYSIS BY PROBABILITY OF OCCURRENCE INDEX

While it is important to analyze the severity index, it is also true that you must analyze the probability of occurrence and level of detectability of potential failures. These three factors are combined to calculate the RPN. Figure B.20 shows the Pareto diagram based on the probability of occurrence of potential failure causes. The scale used to classify the probability of occurrences was as follows:

5 = Always:	More than once per lot
4 = Frequently:	Once per lot
3 = Occasionally:	Once at least every three to six months
2 = Rarely:	Once every one to three years
1 = Never:	Once in more than five years

Notice that the indices are based on probability of occurrences, not on actual number of occurrences. This is an important aspect to consider because the severity, probability of occurrence, and detection indices are independent of each other. In most instances, when the failure occurrence rate is low, it is because good controls are implemented and followed. However, if those controls are removed, the number of occurrences will definitely increase. In order to assign the probability of occurrence, the frequency of the tasks was considered. For example, spillage can occur more than once per lot, because materials are constantly being retrieved and used in the different areas. In these types of situations, a probability of occurrence index of 5 was assigned. However, equipment cleaning happens at the end of the lot. So, for these types of situations, a probability of occurrence index of 4 was assigned.

It should be noted that the scale of probability of occurrence index is highly subjective and not necessarily linear. It is just a subjective way of identifying how often the failure mode could occur. Analyzing Figure B.20, it can be observed that 59.2 percent of the incidents could occur frequently (4), or once per lot, and 40.8 percent of the incidents could occur always (5), which indicates that the cause of the failure could occur more than once per lot.

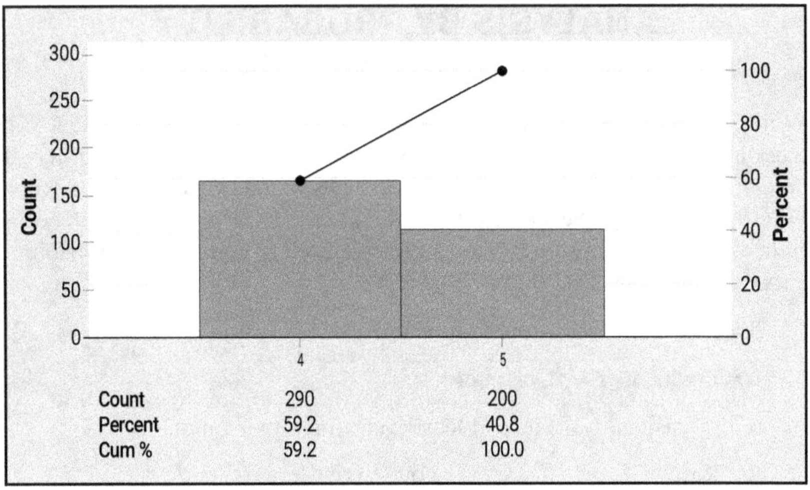

Figure B.20 Pareto diagram by occurrence index.

ANALYSIS BY DETECTABILITY INDEX

Finally, Figure B.21 shows the Pareto diagram classified by level of detectability. The scale used to classify the detection level was the following:

5 = The current control is not capable of preventing or detecting the failure mode.

4 = The action is performed, but there is no procedure defining how it is performed.

3 = A procedure exists, is done, and is registered, but details are missing.

2 = There is a procedure, and the action is performed and recorded.

1 = Control is 100 percent automated.

Analyzing Figure B.21, it can be observed that in 96.7 percent of the situations evaluated in the FMEA, there is a procedure, and the action is performed and recorded (2). In the remaining 3.3 percent of the situations evaluated, the procedure exists, is done, and is registered, but details are missing (3). As mentioned earlier, in order to reduce the RPNs, one of the most important activities is to reduce the detectability index from

3 to 2. It is recommended that XYZ Worldwide Generics Limited evaluate procedures (especially, equipment cleaning procedures) to make them easier to understand. It is also recommended to XYZ Worldwide Generics Limited that the revision of the procedures be linked to a comprehensive training program and adequate supervision.

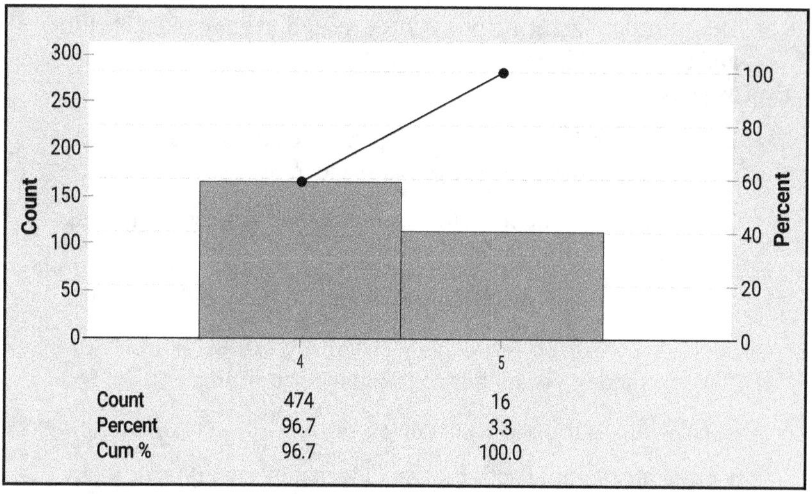

Figure B.21 Pareto diagram by detectability index.

RECOMMENDATIONS

Throughout this case study, a series of recommendations have been offered to be evaluated by XYZ Worldwide Generics Limited. Some other recommendations were previously made after evaluating the QRM procedure. All these recommendations are summarized here:

Recommendations for the QRM procedure:

- Use a systematic approach to identify which tools are going to be used at each stage of the QRM process. Business Excellence Consulting Inc. has developed a Risk Management Toolkit and provided a copy of the toolkit to XYZ Worldwide Generics Limited.
- Use five ratings for each element. Ratings for severity and detectability already defined in procedure can be further improved.

- Business Excellence Consulting Inc. has provided a copy of a QRM model procedure, which is included in Appendix D of this book. It is recommended to implement this procedure entirely or identify the relevant elements and incorporate them in the existing QRM procedure.

Recommendations during the risk assessment:

- Review manufacturing procedures related to equipment cleaning and make certain they have enough details. Use visual aids as much as possible.
- Review the training status of all personnel and make certain they are trained in their tasks, especially cleaning tasks.
- Review training methods to ensure they are aligned to the tasks, especially cleaning tasks.
  - Use a combination of classroom and on-the-job training, as appropriate.
  - Use as many visual aids as possible for training.
  - Evaluate training effectiveness.
- Ensure all cleaning activities are adequately supervised.
- Ensure cleaning activities are double-checked by an independent person.
- Appoint one person as leader of the QRM process oversight.
- Appoint subject matter experts at each area, responsible for the development and maintenance of the FMEAs.
- Implement the QRM procedure model developed by Business Excellence Consulting Inc.

Appendix C
Fail-Safe FMEA

José Rodríguez-Pérez and
Manuel E. Peña-Rodríguez

Quality risk management principles are used effectively in many areas of business and government—including finance, insurance, occupational safety, and public health—and by the agencies regulating these industries.

Risk management's widespread use isn't a surprise because every product and every process has an associated risk. But while there are some examples of the use of quality risk management in the medical product manufacturing industry, they are limited and do not take full advantage of what risk management has to offer. After all, inadequate or ineffective quality risk management can harm patients, product users, and company value.

The current focus of the U.S. Food and Drug Administration (FDA) on risk-based determination requires that regulated industries dramatically improve their understanding and use of hazard control concepts.

The appropriate use of quality risk management can help organizations comply with regulatory requirements, such as good manufacturing practices or good laboratory practices.

Quality risk management is a valuable component of an effective quality system's framework. It can, for example, help guide the setting of specifications and process parameters for manufacturing, assess and mitigate the risk of changing a process or specification, and determine the extent of deviation investigations and corrective actions.

An effective quality risk management approach can ensure a high-quality product by providing a proactive means to identify and control potential quality issues during development and manufacturing. Additionally, it can improve decision making if a quality problem arises.

This article originally appeared in *Quality Progress* magazine, vol. 45, no. 1, January 2012, pp. 30–36.

EXPLAINING RISK

Risk combines the probability of occurrence of harm with the severity of that harm. Quality risk management supports a scientific and practical approach to decision-making during the life cycle of a product. It provides documented and reproducible methods to accomplish the quality risk management process based on current knowledge about the probability, severity, and detectability of the risk.

But each stakeholder perceives different potential harms, computes different probabilities, and assigns different severities. In relation to medical products, although there are many different stakeholders—including patients, medical practitioners, government, and the industry as a whole—protecting the patient by managing risk should always be of the utmost importance.[1]

The manufacture and use of any medical product necessarily involves some degree of risk. The risk to product quality is just one component of overall risk. Product quality must be maintained throughout the product life cycle so the characteristics that are important to product quality remain consistent.

Effective quality risk management can also facilitate better and more informed decisions, assure regulators of a manufacturer's ability to deal with potential risks, and positively affect the extent and level of regulatory oversight.

Manufacturers of regulated products are required to have a quality management system and processes for addressing product-related risks. These processes for managing risk can evolve into a stand-alone management system. While manufacturers may choose to maintain these two management systems separately, it may be advantageous to integrate them because doing so could eliminate redundancies and lead to a more effective management system.

TO EACH THEIR OWN

Risk management is a complex subject because each stakeholder places a different value on the probability of harm occurring and its severity. As one of the stakeholders, the manufacturer makes judgments relating to the safety and performance of a medical product, including the acceptability of risks.

Traditionally, risk is assessed and managed in a variety of informal ways based on a compilation of observations, trends, and other information. That approach can provide useful information that supports the handling of complaints, quality defects, deviations, and resource allocation. But with a more formal approach, industry and regulators can assess and manage risk using recognized risk management tools:

- Basic risk management facilitation methods, such as flowcharts and check sheets
- Failure mode and effects analysis (FMEA)
- Fault tree analysis
- Hazard analysis and critical control points (HACCP)
- Hazard operability analysis
- Preliminary hazard analysis
- Risk ranking and filtering

These tools need to be adapted for specific uses. Quality risk management methods and the supporting statistical tools can be used in combination. The combined use of these tools provides flexibility that can facilitate the application of quality risk management principles.

The degree of rigor and formality of quality risk management should reflect available knowledge and be commensurate with the complexity or criticality of the issue being addressed.

SYSTEMATIC APPROACH

Within pharmaceutical and medical products manufacturing, few tools are more useful than FMEA. In one of its guides for industry—*ICH Q9: Quality Risk Management*—FDA summarizes some of the most common risk management tools. Among the tools mentioned is FMEA.

FDA writes in section I.2, "FMEA provides for an evaluation of potential failure modes for processes and their likely effect on outcomes and/or product performance. Once failure modes are established, risk reduction can be used to eliminate, contain, reduce, or control the potential failures. FMEA relies on product and process understanding. It methodically breaks down the analysis of complex processes into manageable steps. It is a powerful tool for summarizing the important modes of failure, factors causing these failures, and the likely effects of these failures."[2]

Furthermore, it mentions that FMEA can be used to prioritize risks and monitor the effectiveness of risk control activities.[3] The guide defines FMEA and discusses areas of application but does not provide a method to carry out an effective FMEA. For that reason, a systematic approach is needed.

That approach consists of three interrelated tools: a process map, a cause-and-effect matrix (also called a prioritization matrix), and an FMEA. Instead of using the FMEA template from the beginning of the process, let's analyze the failure modes for the inputs that have a larger effect on the critical customer requirements.

Process Map

As opposed to the commonly used flowchart, the process map is a high-level tool that focuses on process steps, as well as their inputs and outputs. One of the main uses of the process map is to identify process inputs that cause high variability.

The process map usually consists of six to 10 boxes that represent major steps in the process. For each step, the inputs and outputs are identified. Figure C.1 shows an example of a partial process map for a tablet packaging facility.

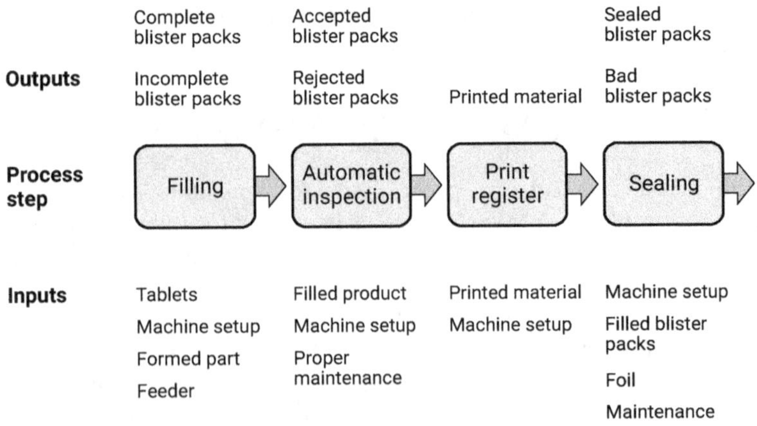

Figure C.1 Partial process map for tablet packaging.

Cause-and-Effect Matrix

After the process map is developed, the prioritization process begins. Many companies start to fill in the FMEA template at this stage. That isn't a good practice because in many processes the most critical steps are located near the middle or toward the end of the process.

Those who use the beginning-to-end approach are likely to be exhausted by the time they start to analyze the critical steps. For that reason, it makes sense to add this intermediate tool.

A cause-and-effect matrix—not to be confused with the cause-and-effect diagram or fishbone—is used to analyze how a process input affects process outputs. To develop a cause-and-effect matrix, follow these six steps:

1. Identify the customer requirements and write them in the upper columns.
2. Assign a rating on a scale with 1 being the least important and 10 the most important. These ratings are assigned from the customer's perspective.
3. Using the process map, list the process steps and process inputs in each row.
4. For each process input, analyze how it affects each customer requirement using the following scale:
    - 0 = The input does not affect the customer requirement.
    - 1 = The input has a hardly noticeable effect on the customer requirement.
    - 3 = The input has a medium effect on the customer requirement.
    - 9 = The input has a strong and direct effect on the customer requirement.
5. Cross-multiply the ratings from steps two and four. For each row, multiply the value of the process input and its corresponding customer requirement. Add those cross-products at the end of each row.
6. Sort the values from step five in descending order.

The end product of using this tool is a list of the process inputs that have the biggest effect on the customer requirements. In this way, you can focus the development of the FMEA on those process inputs that really matter to the customer and leave unimportant issues for later analysis.

Eventually, you should analyze all the process inputs and how they could fail. But the key takeaway is knowing where to focus your efforts. Figure C.2 shows an example of the tablet packaging cause-and-effect matrix already sorted in descending order.

		Rating of importance to customer	10	9	6													
			1	2	3	4	5	6	7	8	9	10	11	12	13	14	15	
			Setup	Rejections	Jam	Requirement	Requirement	Requirement	Requirement	Requirement	Requirement	Requirement	Requirement	Requirement	Requirement	Requirement	Requirement	Total
	Process step	Process input																
4	Filling	Feeder	9	9	9													225
5	Automatic inspection	Filled product	9	9	9													225
7	Print registration	Printed material	9	9	9													225
14	Cartoner	Components	9	9	9													225
15	Cartoner	Setup	9	9	9													225
3	Filling	Setup	9	9	0													171

Figure C.2 Cause-and-effect matrix.

FMEA

After completing the cause-and-effect matrix, you can start to fill in the FMEA template. Instead of the beginning-to-end approach, analyze the failure modes of the inputs that are strongly related to customer requirements.

Using the cause-and-effect matrix, select the process step or input. For that process step or input, consider the possible ways it could fail (the failure mode). Then, for each failure mode, determine the result if it fails (the effect).

After identifying the effects, make a quantitative assessment of the severity. Typical scales range from 1 (no severity) to 10 (critical to human safety). After the severity rating is assigned, identify the potential causes of the failure mode. In this stage, don't think about causal factors or the immediate causes. Instead, focus on potential root causes.

These are labeled as potential causes because causes have not come into play yet. This is one misconception about an FMEA. It's defined as a proactive tool to identify and prevent the way in which a process, product, or system could fail. But in most cases, FMEAs are used after the failure has occurred. That issue will be addressed later.

After you identify the potential root causes, rate the probability of occurrence on a scale from 1 (low probability of occurrence) to 10 (high probability of occurrence). Then determine the current controls used to detect causes or failures and rate their effectiveness on a scale from 1 (controls are 100 percent effective) to 10 (controls are not effective).

At this point, you can calculate the risk priority number (RPN) by multiplying the severity, occurrence, and detectability indices. But that's not the end of the FMEA process. The next step is to sort the table in descending order of RPNs. In that way, the inputs with higher RPNs (most critical to the customer) will be located at the top of the table, while those with lower RPNs (least critical to the customer) will be toward the bottom of the table.

Not sorting RPNs from highest to lowest is another error many organizations make. This is the essence of the FMEA—to prioritize efforts to prevent the most critical inputs to failure. It does not mean you won't manage the least critical inputs; it's a matter of prioritizing.

This is aligned with what FDA expects: that an organization's actions are commensurate with the risk to and impact on patient safety.

After calculating and sorting the RPNs, the real work begins. At this stage, it's possible to identify and implement actions to reduce the RPN. After those actions are implemented, the new RPN must be calculated. This is the fundamental nature of the FMEA process—continuous improvement. Figure C.3 shows an example of an FMEA for a tablet packaging process.

Process step/input	Potential failure mode	Potential failure effects	Severity	Potential causes	Occurrence	Current controls	Detection	RPN	Actions recommended	Responsibility	Actions taken	Severity	Occurrence	Detection	RPN
What is the process step/input under investigation?	In what ways does the key input go wrong?	What is the impact on the key output variables (customer requirements) or internal requirements?	How severe is the effect to the customer?	What causes the key input to go wrong?	How often does cause or failure mode occur?	What are the existing controls and procedures (inspection and test) that prevent either the cause or the failure mode? *Should include an SOP number.*	How well can you detect cause or failure mode?		What are the actions for reducing the occurrence of the cause, or improving detection?	Who's responsible for the recommended action?	What are the completed actions taken with the recalculated RPN? *Be sure to include completion month/year.*				
Filling/ feeder	Excessive vibration	Unfilled cavities	10	Lack of adequate preventive maintenance	8	Automatic inspection	7	560	Include the machine in the PM program	M. Peña	Machine was included in monthly PM program	10	2	3	60
Filling/ feeder	Excessive vibration	Cavities with more than one tablet	10	Lack of adequate preventive maintenance	8	Automatic inspection	7	560	Include the machine in the PM program	M. Peña	Machine was included in monthly PM program				0

RPN = Risk priority number
SOP = Standard operating procedure
PM = Preventive maintenance

Figure C.3 Example of FMEA for tablet packaging.

MAKING THE LINK

At this point, you can integrate the FMEA process with one of the systems used most frequently by an FDA-regulated industry: the corrective action and preventive action (CAPA) system.

Many organizations devote valuable resources to a proactive FMEA. Others focus on the CAPA system to manage the failures within their processes. A select few get the maximum benefit of combining both.

The true preventive actions—actions implemented to eliminate the root causes of nonconformances before they occur—can be easily identified via an effective FMEA initiative. That means you can identify the potential failure modes and their root causes, and then implement the actions required to prevent the occurrence of those causes.

But, in an imperfect world, failures occur. When that happens, an organization immediately activates its CAPA system to initiate an investigation geared to finding the root causes and implementing the appropriate CAPAs.[4] In the vast majority of situations, those systems work independently. But do they need to?

True preventive actions can be identified using the FMEA tool. Then, those actions are incorporated into the preventive action portion of the CAPA system. When a failure does occur, the CAPA system is activated. But because potential failure modes and their causes are part of the FMEA initiative, a disconnect exists. For that reason, the CAPA system should provide feedback to the FMEA tool to revise and update the FMEA accordingly.

FMEA GONE WRONG

Often, organizations face situations in which FDA alerts them of the wrong use of some tools. Here are excerpts from two warning letters issued by FDA regarding the incorrect application of the FMEA tool during the past decade:

1. *Lack of integration of the FMEA process with a performance test.* An organization received customer complaints about its tubing, which was assembled incorrectly. As part of the investigation, the organization conducted a performance test, the result of which showed that the incorrect assembly did not affect the product.

But the organization did not update its FMEA, which stated that the incorrect assembly compromised the functionality of the device.[5] As a result, it left itself vulnerable to future occurrences of the same issue.

2. *Lack of definition of the FMEA indexes for severity, occurrence, and detectability.* An organization received customer complaints about device failures, so it rushed to develop a reactive FMEA based solely on the major cause of complaints: contamination.

When it developed the scales to rank severity, occurrence, and detectability, it did not provide a rationale for each scale. The organization only provided the indexes and the calculated RPN.[6]

These two examples show how the use of FMEA alone isn't enough to guarantee product quality. While FMEA is a proactive tool within an organization's risk management system, it is a dynamic component that must be handled correctly using the following tips:

- Perform an FMEA using a teamwork approach.
- Use a systematic method to complete the FMEA with help from a process map and cause-and-effect matrix.
- Prioritize the order in which the process inputs will be evaluated using the FMEA tool.
- Sort RPNs in descending order.
- Identify all potential root causes.
- Identify and implement CAPAs for each root cause.
- Recalculate RPNs after CAPAs are implemented, not before.

Doing so will ensure that your FMEA is as effective as possible and your products are safe in the hands of the people they're meant to help.

References

1. International Organization for Standardization, *ISO 14971:2007—Medical devices—Application of risk management to medical devices.*
2. U.S. Food and Drug Administration, "Guidance for Industry: Q9 Quality Risk Management," 2006.
3. Ibid.
4. José Rodríguez-Pérez, *CAPA for the FDA-Regulated Industry,* Milwaukee: ASQ Quality Press, 2010.
5. U.S. Food and Drug Administration, "RD Medical Manufacturing Inc. 2/7/11," www.fda.gov/iceci/enforcementactions/warningletters/ucm242822.htm.
6. U.S. Food and Drug Administration, "Vapotherm Inc. 12-Jun-06," www.fda.gov/iceci/enforcementactions/warningletters/2006/ucm075943.htm.

Appendix D

Quality Risk Management Procedure Model

PURPOSE AND SCOPE

1.1 The purpose of this procedure is to define the system for the evaluation of possible risks that may exist at all levels of the organization and facilitate compliance with GMP standards or other quality, safety, and environmental requirements.

1.2 The risk analysis is applied to the development, manufacture, and distribution of the medicinal products that are obtained in ABC Corporation, through their life cycles. The aspects to consider in the risk analysis are the facilities, systems, equipment, or processes of ABC Corporation.

DEFINITIONS

2.1 **Cause:** A defect that is the underlying reason for the failure or that triggers a process that leads to failure.

2.2 **Cause-and-Effects Matrix (CEM):** A work tool used to prioritize system inputs to be evaluated in failure mode and effects analysis.

2.3 **Client:** Internal or external department, people, and/or processes that will be negatively affected by the failure mode. Some examples of clients are: the following operation, the subsequent operations, the end user, etc.

2.4 **Failure Mode and Effects Analysis (FMEA):** An FMEA identifies potential failure modes, determines their possible causes, analyzes associated risks, presents current controls and generates a record of corrective actions.

2.5 **Current Controls:** A mechanism that prevents the cause of the failure mode from occurring or that detects the failure before it reaches the customer.

2.6 **Detectability (D):** The probability that the potential failure mode or its cause can be identified before it reaches the customer.

2.7 **Effect:** An adverse consequence of the failure mode that affects the customer.

2.8 **Failure Mode:** Any way in which a product or process could fail to fulfill its function. A failure mode can be real or potential.

2.9 **Harm:** Damage to health, including damage that may result from the loss of quality of a product or its availability.

2.10 **Hazard:** It is a potential source of harm.

2.11 **Occurrence (O):** Probability that a potential cause will happen during the expected life and the use of the product.

2.12 **Quality Risk Management:** A systematic process for the assessment, control, communication, and review of the risks for the quality of a product (medicines) throughout its life cycle.

2.13 **Risk:** The combination of the probability that a harm will happen and its severity.

2.14 **Risk Acceptance:** The decision to accept the risk.

2.15 **Risk Analysis:** The estimate of the risk associated with an identified problem. It is the process of qualitative or quantitative analysis in which the union between the probability of occurrence and the severity of the damage is estimated. The ability to detect potential damage is a factor that is also used.

2.16 **Risk Assessment:** The systematic process of organization of information that supports decision-making during a process of risk management for quality. It consists in the identification of hazards and in the analysis and evaluation of the risks associated with exposure to such hazards.

2.17 **Risk Communication:** The exchange of information on risk and risk management among decision-makers and other parties.

2.18 **Risk Control:** Actions for the implementation of risk management decisions.

2.19 **Risk Evaluation:** Compares the risk identified and analyzed with a risk criterion. Indicates the tolerance to a risk or its level of acceptability by the organization.

2.20 **Risk Identification:** The systematic use of information to identify potential sources of harm (hazards) concerning the risk in question or the description of the problem.

2.21 **Risk Priority Number (RPN):** The mathematical product of the severity, occurrence, and detection rates (RPN = S × O × D). The RPN is used to prioritize the potential failure modes identified and the consequent corrective and/or preventive actions.

2.22 **Risk Reduction:** Actions taken to reduce both the probability that a harm will occur and the severity of that harm.

2.23 **Risk Revision:** Review or control of the results of the risk management process, taking into account, if applicable, the new knowledge and experience on this risk.

2.24 **Severity (S):** Measure of the possible consequences of a harm.

REFERENCES

3.1 Food and Drug Administration. 2023. *Guidance for Industry: Q9(R1) Quality Risk Management*. Washington, D.C.

3.2 International Society for Pharmaceutical Engineering. 2017. *ISPE Baseline Pharmaceutical Engineering Guide, Volume 7 – Risk-Based Manufacture of Pharmaceutical Products*. Second Edition. Tampa, FL.

3.3 Rodríguez-Pérez, José. 2017. *Quality Risk Management in the FDA-Regulated Industry*. Second Edition. Milwaukee: ASQ Quality Press.

3.4 Rodríguez-Pérez, José, and Peña-Rodríguez, Manuel E. 2012. "Fail-Safe FMEA." *Quality Progress*. January. Milwaukee: ASQ Quality Press.

4. PROCEDURE

4.1 Initiate the Quality Risk Management Process

4.1.1 The systematics of risk analysis will be carried out by a team made up of personnel from the different departments involved. In this team, the quality assurance department will act as coordinator. In general, it can be schematized as shown in Figure D.1:

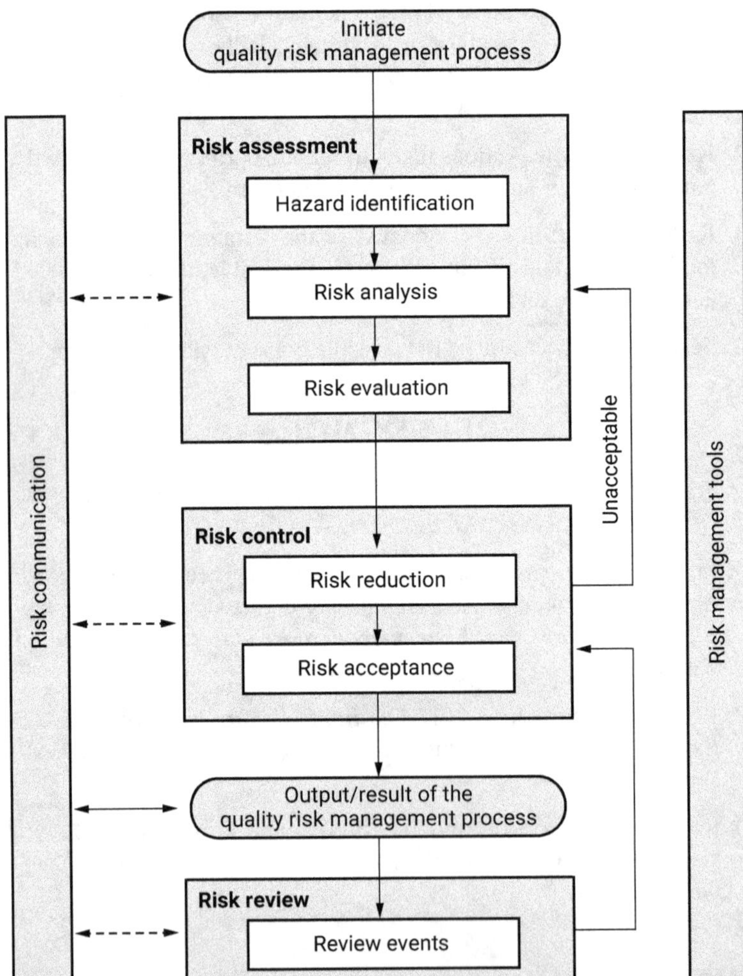

Figure D.1 Quality risk management flowchart.

4.1.2 Quality risk management will include systematic processes designed to coordinate, facilitate, and improve decision-making based on scientific knowledge, with respect to risk.

4.1.3 All risk management will begin with a risk management charter. The charter will include, as a minimum, the following elements:

4.1.3.1 Business case

4.1.3.2 Problem satement

4.1.3.3 Goal statement

4.1.3.4 Project scope

4.1.3.5 Project plan

4.1.3.6 Team

4.1.4 An example of the template for the project charter is presented in Annex 1.

4.2 Risk Assessment

4.2.1 Risk Identification

4.2.1.1 The available information on the system or process to be analyzed will be compiled, so it is defined as clearly as possible.

4.2.1.2 In general terms, the amount and type of information will depend on whether the objective of the risk analysis is on the design or on a machine, process, or system.

4.2.1.3 In the case of process risk analysis, a process map will be established to facilitate its study.

4.2.1.4 A process map is a visual representation of a system in which the main parts or functions are represented by consecutive blocks showing their relationship. The process map is normally used to identify all the inputs of each process, as well as the product of each step of the process.

4.2.1.5 The FMEA team must observe the process and/or equipment while it is in operation to capture the state of the system "as is." It is highly recommended to observe the operation of the system in different conditions, which can be: different shifts, different operators, different products, higher or lower speed of the equipment, etc.

4.2.1.6 Write down the steps as they are carried out. The process can be recorded on video, if this is allowed, to facilitate the identification and evaluation of each step in the process.

4.2.1.7 Develop a process map to visually show the process flow and auxiliary activities (e.g., periodic feed materials, etc.) in a sequential manner.

4.2.1.8 Review the diagram to ensure that all steps of the process are considered.

4.2.1.9 Identify the inputs of each step. An input is what is fed in the step. The inputs can be materials, machinery, methods, measuring equipment, personnel and work environment, among others.

4.2.1.10 An example of the template for the process map is presented in Annex 2.

4.2.1.11 Once the process map has been completed, proceed to identify which of those process inputs are those that most relate to the problem statement established in the project charter.

4.2.1.12 A cause-and-effect matrix (CEM) is used to identify the key process inputs that have the most relationship to the critical-to-quality output variables. This tool offers the initial information for the FMEA. This matrix gives priority to the key inputs that will be analyzed in the FMEA.

4.2.1.13 An example of the template for the CEM is presented in Annex 3 to be used as a working tool each time an AMFE is generated.

4.2.1.14 Start the MCE by writing the critical-to-quality variables (requirement outputs) in the upper cells of each column, as shown in Figure D.2.

4.2.1.15 Assign a rating to each critical-to-quality output variable, based on its importance. The rating process must be carried out by the FMEA team to guarantee a balanced evaluation. Use a scale of 1 to 10, where 1 represents no importance and 10 the most important.

4.2.1.16 List the process steps and the inputs identified in the process map, as shown in Figure D.3.

		Rating of importance to customer														
			1	2	3	4	5	6	7	8	9	10	11	12	13	
Item No.	Process step	Process input	Requirement (Output 1)	Requirement (Output 2)	Requirement (Output 3)	Requirement (Output 4)	Requirement (Output 5)	Requirement (Output 6)	Requirement (Output 7)	Requirement (Output 8)	Requirement (Output 9)	Requirement (Output 10)	Requirement (Output 11)	Requirement (Output 12)	Requirement (Output 13)	Total
1																0
2																0
3																0
4																0
5																0
6																0

Figure D.2 Critical-to-quality output variables in CEM.

4.2.1.17 Assign a rating to each input. These are classified according to their relationship to the output. Use Table D.1 to establish the relationship of the input on each output.

4.2.1.18 The total value, in the column to the right of each row, is calculated as the sum of the products of the rating for each input and the weight of the output; this is:

Total Row 1 = (Input 1 × Output 1) + (Input 1 × Output 2) + (Input 1 × Output 3) + …

		Rating of importance to customer						
			1	2	3	4	5	6
Item No.	Process step	Process input	Requirement (Output 1)	Requirement (Output 2)	Requirement (Output 3)	Requirement (Output 4)	Requirement (Output 5)	Requirement (Output 6)
1								
2								
3								
4								
5								
6								

Figure D.3 Process steps and inputs in CEM.

Table D.1 Input ratings for CEM.

Rating	The input has a ...
9	Strong and direct relationship with critical-to-quality variable
3	Moderate relationship with critical-to-quality variable
1	Remote relationship with critical-to-quality variable
0	No relationship with critical-to-quality variable

4.2.1.19 The total value per row represents the total relationship of each input with respect to the critical-to-quality output variables. Once all the input-output relationships have been calculated, the total values must be ordered in descending order. The rows with the highest total values are those inputs of the process that have the highest relation to the critical-to-quality output variables.

4.2.1.20 The following scale will be used to determine the relationships between the inputs and critical-to-quality output variables:

- For weighted scores that have at least one rating of 9, a red classification is assigned.
- For weighted scores with no rating of 9 but at least one rating of 3, a yellow classification is assigned.
- For weighted scores with no ratings of 9 or 3, but at least one rating of 1, a green classification is assigned.
- For weighted scores with all ratings of zero, a white classification is assigned.

4.2.2 Risk Analysis

4.2.2.1 The development of the FMEA must be carried out by the selected team to ensure that all aspects are taken into consideration and that the risk rating is not biased. An isolated FMEA will not generate an effective analysis.

4.2.2.2 The initial inputs for the FMEA are those identified in the CEM that have the strongest relationship with the critical-to-quality output variables. These are the first elements that are listed in the FMEA format (Annex 4), those with a strong and direct relationship. As mentioned, these are the inputs that have weighted scores with at least one rating of 9.

4.2.2.3 The FMEA team must evaluate each step/input of the system and identify all associated failure modes. A failure mode is a statement of what can go wrong in that particular step and input, that is, ways in which a system can fail potentially to perform its function or the intention of the design. Faults can be partial, intermittent, or unintentional, for example, and can affect quality, reliability, performance, or safety.

4.2.2.4 The result of a failure mode that is not corrected, detected, or removed is its effect. The effect of a failure mode is the impact it will have on the client (requirements/outputs) and describes in terms of the client's perception. It can be an immediate effect in the subsequent stage or beyond in the life of the product.

4.2.2.5 The severity of each effect is the seriousness of this effect for the client. This is a relationship with an ordinary value in relation to the harm that may be caused by the effect of the patient's risk, the regulatory impact, and/or the impact of the process. Annex 5 shows the severity rating in the FMEA.

4.2.2.6 Identify the potential causes that could lead to the failure mode under study. A failure mode can have more than one cause; however, the analysis focuses on possible root causes based on exhaustive assessments rather than on the causal factors or immediate causes.

4.2.2.7 A cause-and-effect diagram (fishbone diagram) and a fault tree analysis are useful tools that can be used to identify potential causes. "Incorrect material" or "Lack of a standard operating procedure" are examples of potential causes.

4.2.2.8 The causes are classified according to their probability of occurrence. The probability of a failure that occurs during the operation of an identified cause is indicated by an ordinal value scale that ranges from 1 (will never happen) to 5 (will always happen). Annex 6 shows the occurrence rating in the FMEA.

NOTE: The probability of occurrence is not based on the actual number of occurrences the failure has happened, but on the potential that the failure occurs every time a task is performed.

4.2.2.9 The column of controls in the FMEA is filled by a list of current methods that prevent a failure mode from occurring (proactive) and those that detect the failure mode if it occurs (reactive).

4.2.2.10 Conduct an assessment of the likelihood that the current controls will detect the cause or failure mode before the product leaves the current process stage, the installation, or before its delivery to the client.

4.2.2.11 The rating of the efficacy of the controls is established by the detection value. This value represents the capacity to identify the presence of a failure mode at the earliest possible point in the process.

4.2.2.12 An easily detected failure mode, given the current controls, is assigned a low value, while a high value indicates that the detection of a failure mode is very unlikely. Annex 7 shows the detection rating in the FMEA.

4.2.2.13 The priority risk number (RPN) can be calculated once the severity, occurrence, and detection have been determined. The RPN is the measure of the risk of failure and is used to prioritize corrective/preventive actions. Mathematically, the RPN is the product of severity, occurrence, and detection.

RPN = Severity x Occurrence x Detection

4.2.3 Risk Evaluation

4.2.3.1 Classify the failure modes (rows) in descending order by the RPN so the rows with the highest RPN (which represents the highest risk to the client) are at the top of the analysis.

4.2.3.2 The following scale will be used to determine the risk of each of the failure modes:

- 60-125 – High risk
- 27-59 – Moderate risk
- 1-26 – Low risk

NOTE: Regardless of the RPN value, any case with a Severity Index 5 will be considered high risk. In this case, further controls to avoid the failure mode will be analyzed.

4.2.3.2 Once the most significant risks have been identified using the RPN, then actions to reduce severity, occurrence, and/or detection ratings can be recommended.

4.2.3.3 In general, the severity of a failure mode can only be reduced by design changes.

4.2.3.4 The occurrence can be reduced by eliminating or controlling the causes within the process. Automated systems that prevent failure from occurring are a common approach to this classification.

4.2.3.5 Detecting a failure mode is usually the first way to respond to reduce the RPN. Improving existing control methods or implementing new ones, such as automated inspection systems and online tests, are actions that allow the failure mode to be detected before it reaches the client.

4.2.3.6 Assign a responsible person for each recommended action based on the nature of the task.

4.2.3.7 Once the actions are implemented in response to the failure modes, recalculate the RPN taking into account the changes and sort again in descending order.

4.2.3.8 The revision of the FMEA is done either because of the changes implemented that impacted the system or at least on an annual basis. The revision and creation of FMEA documents must be managed by the applicable change control procedures.

4.2.3.9 The FMEA must be updated each time a new failure mode, cause, and/or effect is identified. The identification may be the result of a validation/qualification, case of noncompliance, product/process changes, complaints, or as part of a CAPA activity.

4.3 Risk Control

4.3.1 Risk Reduction

4.3.1.1 Risk reduction is based on processes that mitigate or avoid a quality risk when it exceeds a specific (acceptable) level (see Figure D.1).

4.3.1.2 The risk reduction can include actions that mitigate the severity and probability of the harm. Processes that improve the ability to detect hazards and the risks to quality can also be used as part of the risk control strategy.

4.3.1.3 The implementation of measures to reduce risks can introduce new risks into the system or increase the importance of other existing risks. Hence, it is appropriate to return to the risk assessment to identify and assess any possible change in risk once the process of reducing it has been implemented.

4.3.2 Risk Acceptance

4.3.2.1 The acceptance of the risk is the decision to assume the risk.

4.3.2.2 The acceptance of the risk may be a formal decision in which the residual risk is assumed, or it may be a passive decision in which residual risks are not specified.

4.3.2.3 In some cases, even the best risk management process for quality could not completely eliminate the risk. In these cases, it could be accepted that an adequate strategy has been applied to manage the quality risk and that the risk has been reduced to a specific (acceptable) level.

4.3.2.4 This level will depend on different parameters and should be determined on a case-by-case basis.

4.4 Risk Communication

4.4.1 Risk communication involves the exchange of information about risk and its management among decision-makers and other parties.

4.4.2 Communication can be done at any point in the risk management process. The results of this process must be communicated and documented in the appropriate form.

4.4.3 Communications can be made between different stakeholders, for example, regulators and industry, industry and patients, people within a company, industry and regulatory authority, etc.

4.4.4 The information that is transmitted may be the existence, nature, form, probability, severity, degree of acceptance, control, treatment, detection capacity, or other aspects related to quality risks.

4.4.5 The communication does not have to be done for each and every one of the acceptances of a risk.

4.4.6 The communication between the industry and the regulatory authorities on the risk management decisions for quality could be made through the channels specified in the regulations and guidelines.

4.5 Risk Review

4.5.1 Risk management must be something continuous in the process of quality management.

4.5.2 A mechanism to review and control events must be implemented.

4.5.3 The results of the risk management process should be reviewed taking into account new knowledge and experience.

4.5.4 Once a risk management process for quality has been initiated, it should continue to be used on events that could have an impact on the original decision adopted, whether these events were planned (for example, results after the review of products, inspections, audits, exchange control, etc.) or not (for example, withdrawals, causes found after the investigations of failures, etc.)

4.5.5 The frequency of the review will depend on the level of risk.

4.5.6 The risk review could include a reconsideration of the decisions taken to accept the risk.

4.5.7 It must be kept in mind that the introduction of actions is a possible source of new dangers, so it will be necessary to follow up in order to identify and control them.

4.5.8 The frequency of review will depend on the level of risk and will be established on a case-by-case basis.

5. DOCUMENTS

In each risk analysis, the following must be documented:

- Risk management charter
- Process map
- Cause-and-effects matrix
- Failure mode and effects analysis (FMEA)
- Corrective and/or preventive actions from the FMEA

Code:

Each document will have an assigned code:

RA-ZZZZZZZZ-MMYYYY

whereas:

RA means risk analysis

ZZZZZZZZ are letters or numbers that identify the system or process to be evaluated (maximum eight letters).

MMYYYY are the month (MM) and the year (YYYY) in which the risk assessment was developed.

In case several risk analysis studies are carried out on the same element, these will be distinguished by a consecutive extra number that will be added to the end of the code after a point (starting with 1). In this way, it is understood that the initial report is zero (even if it does not appear in the code).

Risk management title:	
Date (last revision):	
Prepared by:	
Approved by:	

<table>
<tr><th colspan="4">Business Case:</th><th colspan="2">Problem Statement:</th></tr>
<tr><td colspan="4"></td><td colspan="2"></td></tr>
<tr><th colspan="4">Goal Statement:</th><th colspan="2">Project Scope:</th></tr>
<tr><td colspan="4" rowspan="2"></td><td colspan="2">Process start point:</td></tr>
<tr><td colspan="2">Process end point:</td></tr>
<tr><td colspan="4" rowspan="2">Expected benefits:</td><td colspan="2">In scope:</td></tr>
<tr><td colspan="2">Out of scope:</td></tr>
<tr><th colspan="4">Project Plan:</th><th colspan="2">Team:</th></tr>
<tr><th>Task/phase</th><th>Start date</th><th>End date</th><th>Actual end</th><th>Name</th><th>Role</th></tr>
<tr><td></td><td></td><td></td><td></td><td></td><td></td></tr>
<tr><td></td><td></td><td></td><td></td><td></td><td></td></tr>
<tr><td></td><td></td><td></td><td></td><td></td><td></td></tr>
<tr><td></td><td></td><td></td><td></td><td></td><td></td></tr>
<tr><td></td><td></td><td></td><td></td><td></td><td></td></tr>
<tr><td></td><td></td><td></td><td></td><td></td><td></td></tr>
<tr><td></td><td></td><td></td><td></td><td></td><td></td></tr>
</table>

Annex 1 Risk management charter template.

Annex 2 Process map template.

Process/project name:	
Date:	
Performed by:	
Notes:	

Output Variables (Y's):				
Importance Score (1–10):				

Process Step	Process Input	Table of Association Scores (X's to Y's)				Weighted Score
						0
						0
						0
						0
						0
						0
						0
						0
						0
						0
						0
						0
						0

Annex 3 Cause-and-effect matrix template.

Process/product:		FEMA Date: (original):	
FMEA team:		(Revised):	
Responsibility:		Page:	of
Prepared by:			

Row No.	Process										Actions			Results			
	Process Step	Process Input	Potential Failure Mode	Potential Effects of Failure	Severity	Potential Cause(s) of Failure	Occur-rence	Current Controls	Detection	Risk Priority Number (RPN)	Recommended Action	Responsibility and Target Completion Date	Action Taken	Revised Severity	Revised Occurrence	Revised Detection	Revised Risk Priority Number

Annex 4 Failure mode and effects analysis (FMEA) template.

Rating	Severity
5	**Critical**—The performance and safety of the product are severely affected, and the client is very dissatisfied. It endangers the machinery or the operator. The product is not functional. It leads to noncompliance of a regulation and/or safety specifications, as well as product recall, consequences to health, or death. Causes injury to a patient or another person.
4	**Major**—Product performance is affected to a large extent, and the client is very dissatisfied. It can cause serious disruption to subsequent operations or require further rework. The product may become inoperable, but it remains safe. Field alert or several affected lots.
3	**Moderate**—The product has a minor violation and is not subject to legal actions of regulatory agency. There is a moderate effect on product performance. The client experiences some dissatisfaction. It may cause rework, repair, or damage to the equipment. The product is operable but is something that can result in a customer complaint.
2	**Minor**—There is a slight effect on product performance. The client may be a little annoyed. It may cause slight discomfort with subsequent operations. Failure does not require product rework or repair; rather, it is a small annoyance that does not cause loss.
1	**No Severity**—There is no effect or very little effect, that is, no consequences, adequate product in adequate quantity in the correct container, correctly labeled. The process is in control. The client probably will not even notice the effect of this failure mode on the product. It is not perceived and does not affect performance.

Annex 5 Severity ratings for FMEA.

Rating	Occurrence
5	**Always—**The failure is almost sure to happen. The failure can occur more than once a day and is almost certain to occur more than once per batch.
4	**Frequently**—The failure will occur in most circumstances. The failure can occur once a week, once per month, or once per batch.
3	**Moderately**—The failure could happen at any time. There is a moderate chance the cause will occur at least once every three to six months.
2	**Rarely**—The failure is not likely to occur. There is a remote likelihood that the cause will occur at least once every one to three years.
1	**Never**—The failure is extremely unlikely or can never occur. There is a very remote likelihood the cause will occur at least once in more than five years.

Annex 6 Occurrence ratings for FMEA.

Rating	Occurrence
5	**Never**—There are no current controls; they are unreliable or not tested in the detection of this potential cause or failure mode. There is very high probability that the failure will not be detected in time to avoid the effect. The ruling is latent, it is not verified at present, or it is not verifiable. It is not detectable by current methods. The current control is not able to prevent or detect the failure mode.
4	**Rarely**—It is very likely that the failure will not be detected in the process. All inspection is manual. The action is performed, but there is no standard operating procedure defining how it is performed.
3	**Frequently**—The failure may not be detected in the process. Standard operating procedure exists and is done and registered, but details of the standard operating procedure are missing. Standard operating procedure needs improvement.
2	**Almost Always**—It is very likely that the fault will be detected by the control in the process. There is a standard operating procedure, and the action is performed and recorded. Standard operating procedure does not need any improvement.
1	**Always**—The control will detect or prevent failure in almost all cases in the process. Failure is obvious or 100 percent detected. Control is 100 percent automated.

Annex 7 Detectability ratings for FMEA.

Appendix E
Useful Websites

http://ec.europa.eu/health/index_en.htm
The body of European Union legislation in the medical product sector is compiled here.

http://www.asq.org
The American Society for Quality (ASQ) is the world's leading membership organization devoted to quality. This site provides useful information, resources, and links for quality topics.

http://www.fda.gov
This is the entry page to the U.S. Food and Drug Administration website.

http://www.fda.gov/ora
This page contains significant ORA (Office of Regulatory Affairs) documents (consent decrees, 483 forms, Establishment Inspection Reports, and many more regulatory documents) under its ORA FOIA Electronic Reading Room.

http://www.fda.gov/ICECI/EnforcementActions/WarningLetters/default.htm
This is the place to see FDA-published Warning Letters sent to regulated firms.

http://www.fda.gov/Safety/Recalls/default.htm
This section includes the most significant product actions over the last five years based on the extent of distribution and the degree of health risk. In this section, you will find a listing of FDA and industry press releases regarding product recalls. It includes a link to weekly FDA Enforcement Reports.

http://www.bec-global.com
This is the author's page devoted to general quality topics.

http://www.ich.org
The International Conference on Harmonization of Technical Requirements for Registration of Pharmaceuticals for Human Use (ICH) is a unique project that brings together the regulatory authorities of Europe, Japan, and the United States, and experts from the pharmaceutical industry in the three regions, to discuss scientific and technical aspects of product registration.

http://www.ghtf.org
The Global Harmonization Task Force (GHTF) was conceived in 1992 in an effort to achieve greater uniformity between national medical device regulatory systems. A partnership between regulatory authorities and regulated industry, the GHTF is made up of five founding members: European Union, United States, Canada, Australia, and Japan. This webpage will redirect you to http://www.imdrf.org.

http://www.imdrf.org
The International Medical Device Regulators Forum (IMDRF) was conceived in February 2011 as a forum for discussing future directions in medical device regulatory harmonization. It is a voluntary group of medical device regulators from around the world that have come together to build on the strong foundational work of the Global Harmonization Task Force on Medical Devices (GHTF) and to accelerate international medical device regulatory harmonization and convergence.

http://www.iso.org
This is the page of the International Organization for Standardization (ISO).

http://www.hc-sc.gc.ca/dhp-mps/md-im/applic-demande/guide-ld/index_e.html
This page contains the Canadian guidances for the interpretation of the classification rules for medical devices.

Glossary

C

continuous process verification—An alternative approach to process validation in which manufacturing process performance is continuously monitored and evaluated.

D

Decision-maker(s)—Person(s) with the competence and authority to make appropriate and timely quality risk management decisions.

detectability—The ability to discover or determine the existence, presence, or fact of a hazard.

H

harm—Physical injury or damage to the health of people, or damage to property or the environment. Also, damage to health, including the damage that can occur from loss of product quality or availability.

hazard—Potential source of harm.

P

process analytical technology (PAT)—A system for designing, analyzing, and controlling manufacturing through timely measurements (that is, during processing) of critical quality and performance attributes of raw and in-process materials and processes with the goal of ensuring final product quality.

process robustness—Ability of a process to tolerate variability of materials and changes of the process and equipment without negative impact on quality.

product life cycle—All phases in the life of a product from the initial development through marketing until the product's discontinuation.

Q

quality—The degree to which a set of inherent properties of a product, system, or process fulfills requirements.

quality risk management—A systematic process for the assessment, control, communication, and review of risks to the quality of a drug product across the product life cycle.

quality system—The sum of all aspects of a system that implements quality policy and ensures that quality objectives are met.

R

requirements—The explicit or implicit needs or expectations of the patients or their surrogates, for example, healthcare professionals, regulators, and legislators.

residual risk—Risk remaining after protective measures have been taken.

risk—Combination of the probability of occurrence of harm and the severity of that harm.

risk acceptance—The decision to accept risk.

risk analysis—Systematic use of available information to identify hazards and estimate risk.

risk assessment—Overall process comprising a risk analysis and a risk evaluation.

risk communication—The sharing of information about risk and risk management between the decision-maker and other stakeholders.

risk control—Actions implementing risk management decisions (ISO Guide 73).

risk evaluation—Judgment, on the basis of risk analysis, of whether a risk that is acceptable has been achieved in a given context based on the current values of society.

risk identification—The systematic use of information to identify potential sources of harm (hazards) referring to the risk question or problem description.

risk management—The systematic application of quality management policies, procedures, and practices to the tasks of assessing, controlling, communicating, and reviewing risk.

risk reduction—Actions taken to lessen the probability of occurrence of harm and the severity of that harm.

risk review—Review or monitoring of output/results of the risk management process considering (if appropriate) new knowledge and experience about the risk.

S

severity—A measure of the possible consequences of a hazard.

stakeholder—Any individual, group, or organization that can affect, be affected by, or perceive itself to be affected by a risk. Decision-makers might also be stakeholders. It can be the patient, healthcare professional, regulatory authority, and industry.

T

trend—A statistical term referring to the direction or rate of change of a variable(s).

Bibliography

Ahmed, Ruhi, Harold Baseman, Jorge Ferreira, Thomas Genova, William Harclerode, Jeffery Hartman, Samuel Kim, Nanette Londeree, Michael Long, William Miele, Timothy Ramjit, Marlene Raschiatore, and Charles Tomonto. 2008. "PDA Survey of Quality Risk Management Practices in the Pharmaceutical, Devices, & Biotechnology Industries." PDA's Risk Management Task Force. *PDA Journal of Pharmaceutical Science and Technology* 62 (1): 1–21.

Ballentine, Carol. 1981. "Taste of Raspberries, Taste of Death: The 1937 Elixir Sulfanilamide Incident." *FDA Consumer,* June. https://www.fda.gov/files/about%20fda/published/The-Sulfanilamide-Disaster.pdf

Beck, Gregory, Mark Schenerman, John Dougherty, Ruth Cordoba-Rodriguez, Christopher Joneckis, Anthony Mire-Sluis, and Lorna D. McLeod. 2009. "Raw Material Control Strategies for Bioprocesses." *BioProcess International,* September, 18–33.

Bills, Edwin, Stan Mastrangelo, and Fubin Wu. 2015. "Documenting Medical Device Risk Management through the Risk Traceability Summary." *Horizons*, Spring, 26–33

Bogdanich, Walt, and Jake Hooker. 2007. "From China to Panama, a Trail of Poisoned Medicine." *The New York Times*. http://www.nytimes.com/2007/05/06/world/americas/06poison.html.

Daniel, Amiram, and Ed Kimmelman. 2008. *The FDA and Worldwide Quality System Requirements Guidebook for Medical Devices,* 2nd ed. Milwaukee: ASQ Quality Press.

Note: All FDA guidances can be downloaded from http://https://www.fda.gov/regulatory-information/search-fda-guidance-documents.

European Commission,. 2015. *EudraLex Volume 4 EU Guidelines for Good Manufacturing Practice for Medicinal Products for Human and Veterinary Use Annex 15: Qualification and Validation*. Brussels: EC.

Falce, Luca. 2015. "Risk Analysis and Annual Training Program Definition." *Pharmaceutical Engineering*, January–February, 44–52.

Fetterman, Jeffrey E., Wayne L. Pines, and Gary H. Slatko. 2008. *Pharmaceutical Risk Management: Practical Applications*. Washington D.C.: Food and Drug Law Institute.

Food and Drug Administration. 1978a. Code of Federal Regulations Title 21 Part 210: *Current Good Manufacturing Practice in Manufacturing, Processing, Packing, or Holding of Drugs; General*. http://www.accessdata.fda.gov/scripts/cdrh/cfdocs/cfcfr/CFRSearch.cfm?CFRPart=210.

———. 1978b. Code of Federal Regulations Title 21 Part 211: *Current Good Manufacturing Practice for Finished Pharmaceuticals*. http://www.accessdata.fda.gov/scripts/cdrh/cfdocs/cfcfr/CFRSearch.cfm?CFRPart=211.

Preamble of finished pharmaceuticals can be downloaded from https://www.fda.gov/files/drugs/published/Federal-Register-43-FR-45077.pdf

———. 1996. Code of Federal Regulations Title 21, Subchapter H— Medical Devices Part 820: *Quality System Regulation*. http://www.accessdata.fda.gov/scripts/cdrh/cfdocs/cfcfr/CFRSearch.cfm?CFRPart=820.

Preamble of Medical Devices *Quality System Regulation* can be downloaded from https://www.fda.gov/medical-devices/quality-system-qs-regulationmedical-device-good-manufacturing-practices/medical-devices-current-good-manufacturing-practice-cgmp-final-rule-quality-system-regulation

———. 1999a. *Guide to Inspections of Quality Systems*. Washington, D.C.: FDA.

———. 1999b. *Managing the Risks from Medical Product Use: Creating a Risk Management Framework*. Report to the FDA Commissioner from the Task Force on Risk Management. Washington, D.C.: FDA.

———. 2002. *General Principles of Software Validation; Final Guidance for Industry and FDA Staff*. Washington, D.C.: FDA.

———. 2003. *Guidance for Industry Part 11, Electronic Records; Electronic Signatures—Scope and Application*. Washington, D.C.: FDA.

———. 2004a. *Guidance for Industry: Sterile Drug Products Produced by Aseptic Processing—Current Good Manufacturing Practice*. Washington, D.C.: FDA.

———. 2004b. *Risk-Based Method for Prioritizing CGMP Inspections of Pharmaceutical Manufacturing Sites—A Pilot Risk Ranking Model*. Washington, D.C.: FDA.

———. 2006. *Guidance for Industry: Quality Systems Approach to Pharmaceutical CGMP Regulations*. Washington, D.C.: FDA.

———. 2009a. *FDA's Approach to Medical Product Supply Chain Safety. Report to Committee on Appropriations* (U.S. Congress). Washington, D.C.: FDA.

———. 2009b. *Guidance for Industry: Q8(R2) Pharmaceutical Development.* Washington, D.C.: FDA.

———. 2011a. *Guidance for Industry: Process Validation: General Principles and Practices.* Washington, D.C.: FDA.

———. 2011b. *Guidance for Industry: Oversight of Clinical Investigations—A Risk-Based Approach to Monitoring.* Washington, D.C.: FDA.

———. 2012. *Guidance for Industry: Heparin for Drug and Medical Device Use: Monitoring Crude Heparin for Quality.* Washington, D.C.: FDA.

———. 2015a. *Guidance for Industry: Premarketing Risk Assessment (Premarketing Guidance.)* Washington, D.C.: FDA.

———. 2015b. *Guidance for Industry: Development and Use of Risk Minimization Action Plans (RiskMAP Guidance).* Washington, D.C.: FDA.

———. 2015c. *Guidance for Industry: Good Pharmacovigilance Practices and Pharmacoepidemiologic Assessment (Pharmacovigilance Guidance).* Washington, D.C.: FDA.

———. 2015d. *Guidance for Industry: Balancing Premarket and Postmarket Data Collection for Devices Subject to Premarket Approval.* Washington, D.C.: FDA.

———. 2016a. *Draft Guidance for Industry Data Integrity and Compliance With CGMP Guidance for Industry.* Washington, D.C.: FDA.

———. 2016b. *Draft Guidance for Industry Factors to Consider Regarding Benefit-Risk in Medical Device Product Availability, Compliance, and Enforcement Decisions.* Washington, D.C.: FDA.

———. 2016c. *Guidance for Industry and Food and Drug Administration Staff—Factors to Consider When Making Benefit–Risk Determinations in Medical Device Premarket Approvals and De Novo Classifications.* Washington, D.C.: FDA.

———. 2018. *Guidance for Industry: Data Integrity and Compliance With Drug CGMP: Questions and Answers.* Washington, D.C.: FDA.

———. 2022. *Draft Guidance for Industry: FDA Risk Management Plans to Mitigate the Potential for Drug Shortages.* Washington, D.C.: FDA.

———. 2023a. *Guidance for Industry: Q9(R1) Quality Risk Management.* Washington, D.C.: FDA.

———. 2023b. "Enforcement Reports." Recalls, Market Withdrawals, & Safety Alerts. https://www.fda.gov/safety/recalls-market-withdrawals-safety-alerts/enforcement-reports

———. 2023c. "Recalls, Corrections and Removals (Devices)." Postmarket Requirements (Medical Devices). https://www.fda.gov/medical-devices/postmarket-requirements-devices/recalls-corrections-and-removals-devices

———. 2023d. "FDA Food Safety Modernization Act (FSMA)." https://www.fda.gov/food/guidance-regulation-food-and-dietary-supplements/food-safety-modernization-act-fsma

———. 2023e. Medication Guides. Drug Safety and Availability. https://www.fda.gov/drugs/drug-safety-and-availability

Food and Drug Administration, Center for Devices and Radiological Health. 1996. *Do It By Design: An Introduction to Human Factors in Medical Devices*. Washington, D.C.: FDA.

Frank, Ted, Stephen Brooks, Kristin Murray, Steve Reich, Ed Sanchez, Brian Hasselbalch, Kwame Obeng, and Richard Creekmore. 2011. "Quality Risk-Management Principles and PQRI Case Studies." *Pharmaceutical Technology* 35 (7): 72–76.

Global Harmonization Task Force. 2005. GHTF/SG3/N15R8 *Implementation of Risk Management Principles and Activities within a Quality Management System*. GHTF. https://www.imdrf.org/documents/ghtf-final-documents/ghtf-study-group-3-quality-systems

Hauser, Robert G., and Barry J. Maron. 2005. "Lessons from the Failure and Recall of an Implantable Cardioverter-Defibrillator." *Circulation* 112 (13): 2040–42

Health Canada. 2022. *Guidance on Risk Classification of Medical Device Observations (GUI-0079)*. https://www.canada.ca/en/health-canada/services/drugs-health-products/compliance-enforcement/information-health-product/medical-devices/guidance-risk-classification-medical-device-observations-0079.html

International Conference on Harmonization (ICH). 2010. Q8(R2): *Pharmaceutical Development*. https://ich.org/page/training-programme-q8q9q10-0

International Electrotechnical Commission. 2006. IEC 60812 *Analysis techniques for system reliability—Procedure for failure mode and effects analysis (FMEA)*. Geneva: IEC.

International Organization for Standardization (ISO). 2009. ISO Guide 73:2009 *Risk management—Vocabulary*. Geneva: ISO.

———. 2015. ISO 9001:2015 *Quality management systems—Requirements*. Geneva: ISO.

———. 2016. ISO 13485:2016 *Medical devices—Quality management systems—Requirements for regulatory purposes*. Geneva: ISO.

———. 2018a. ISO 22000: 2018 *Food safety management systems—Requirements for any organization in the food chain*. Geneva: ISO.

———. 2018b. ISO 31000:2018 *Risk Management—Guidelines*. Geneva: ISO.

———. 2019a. ISO 14971:2019 *Medical devices—Application of risk management to medical devices*. Geneva: ISO.

———. 2019b. ISO 31010:2019 *Risk Management—Risk assessment techniques*. Geneva: ISO.

______. 2020. ISO/TR 24971:2020 *Medical devices—Guidance on the application of ISO 14971*. Geneva: ISO.

Leveson, Nancy. 1995. "Medical Devices: The Therac 25." In *Safeware: System Safety and Computers*. Boston: Addison-Wesley.

Linstone, Harold A., and Turoff, Murray, eds. 1975. *The Delphi Method: Techniques and Applications*. Boston: Addison-Wesley.

Makary, Martin, and Michael Daniel. 2016. "Medical Error—the Third Leading Cause of Death in the U.S." *The British Medical Journal*. http://www.bmj.com/content/353/bmj.i2139.

Parenteral Drug Association. 2008. Technical Report No. 44, *Quality Risk Management for Aseptic Processes*. Supplement Volume 62, No. S-1. Bethesda, MD: PDA.

Peña-Rodríguez, Manuel E. 2019. *Process Monitoring and Improvement Handbook*, Second Edition. Milwaukee: ASQ Quality Press.

Rodríguez-Pérez, José. 2022. *Handbook of Investigation and Effective CAPA Systems*, Third Edition. Milwaukee: ASQ Quality Press.

Rodríguez-Pérez, José. 2023. *Human Error Reduction in Manufacturing*. Second Edition. Milwaukee: ASQ Quality Press.

Rodríguez-Pérez, José, and Manuel E. Peña-Rodríguez. 2012. "Fail-Safe FMEA." *Quality Progress*, January, 31–36.

Tague, Nancy R. 2005. *The Quality Toolbox*, Second Edition. Milwaukee: ASQ Quality Press.

White, Ed. 2009. "Risk Management for Aseptic Processing." *The Journal of Validation Technology* 15 (2): 25–34.

World Health Organization. 2003. "Application of Hazard Analysis and Critical Control Point (HACCP) Methodology to Pharmaceuticals" (Annex 7). In WHO Technical Report Series No. 908, *WHO Expert Committee on Specifications for Pharmaceutical Preparations*, 37th report. Geneva: WHO.

———. 2015. "Food Safety." Fact sheet No. 399. WHO Media Centre. http://www.who.int/mediacentre/factsheets/fs399/en/.

www.ingramcontent.com/pod-product-compliance
Lightning Source LLC
LaVergne TN
LVHW052038200926
848481LV00005B/1721